THE COMPLETE ILLUSTRATED GUIDE TO FARMING

Samantha Johnson and
Philip Hasheider

Published in 2014 by Voyageur Press, an imprint of Quarto Publishing Group USA Inc., 400 First Avenue North, Suite 400, Minneapolis, MN 55401 USA

ISBN-13: 978-0-7603-4555-9

Library of Congress Cataloging-in-Publication Data

Johnson, Samantha.
The complete illustrated guide to farming / by Samantha Johnson and Philip Hasheider.
pages cm
Other title: Farming
Includes index.
ISBN 978-0-7603-4555-9 (softbound)
1. Agriculture. 2. Farms, Small. I. Hasheider, Philip, 1951- II. Title. III. Title: Farming.
S496.J64 2014
630--dc23

2013047866

Front cover photo: Photo collage, all files from Shutterstock. goats (back): grafvsion, goats (front): oorka, cows: smereka, poultry: PCHT, barn: Thomas Barrat

Back cover photos: (top)Rick Kubik; (bottom) Fotokostic/ shutterstock

Printed in China
10 9 8 7 6 5 4 3 2 1

Design Manager: James Kegley
Layout design: Helena Shimizu
Cover designer: Gavin Duffy
Editor: Elizabeth Noll

NOTE: Many activities described in this book are dangerous. Failure to follow safety procedures may result in serious injury or death. The publisher cannot assume responsibility for any damage to property or injury to persons as a result of misuse of the information provided.

Dedication

To our farm friends and neighbors in Phelps, who are always there to lend a hand: Thanks!
Samantha Johnson

To Jerry Apps, a longtime friend and mentor: his many writings about agricultural subjects and history have served as an inspiration for me.
Philip Hasheider

Contents

Chapter 6: Harvest, Preserve, and Butcher

Chapter 7: Conservation

Introduction

As children, we learn—courtesy of colorful picture books—that farmers wear overalls and a cap, drive green (or sometimes red) tractors, and have cattle, sheep, pigs, and a rooster. We learn that farmers have golden fields of wheat and big red barns.

But as time goes on, we discover that farming can mean many different things. Some farmers raise horses, some keep rabbits, some raise bees. On any given day, a farmer might harvest fresh vegetables from the garden, bottle honey, haul hay, or deliver a newborn foal. Some farmers invest in brand-new machinery while others have assorted tractors of varying degrees of antiquity. Some approach farming with an eye on turning a profit; others focus on the self-sustaining aspects of growing and preserving food for personal use.

You might ask yourself: why farm? But the more pertinent question is: why not farm? Why not experience the deep, fulfilling joy that comes from harvesting honey, raising chicks, growing wheat, and making pickles? Why not pursue these satisfying skills and the accompanying rewards?

That's not to say that the work isn't hard—sometimes it can be. Farming is repetitive in a rhythmic, cyclical way that is by turns

comforting and frustrating; some people find it invigorating, while others flee for the city and never look back.

I farm because it is a way of life; because there are innumerable benefits—both tangible and intangible—and moments that are too precious to miss. The birth of a foal, the moment when a plant sprouts and springs from the soil, the taste of a sun-ripened tomato, the thrilling sight of hay bales stacked neatly in the barn.

Maybe farming knowledge wasn't passed down to you from a previous generation. Maybe you don't have a nearby mentor to answer questions or offer advice. Maybe you're not sure if you want to start out with cattle, or chickens, or rabbits. Maybe you don't want to keep livestock at all. Maybe you think you don't have a green thumb and are scared to attempt to grow your own vegetables. Maybe you'd like to trying canning and preserving, but you aren't sure where to start.

So you've picked up this book. As the title, *The Complete Illustrated Guide to Farming*, suggests, it covers a vast array of topics, and it's fully illustrated to present the material in the most effective and helpful manner possible. You'll find extensive information on livestock, crops, gardening, harvesting, and more, along with resources and tips to help you get started or continue in the farming lifestyle.

Are you ready to take the plunge into "real farming"? Whether your definition includes chickens, goats, wheat, honeybees, or any combination thereof, keep this book by your side and press on—wondrous adventures await.

—Samantha Johnson

What does it take to be a successful farmer, whether large or small, regardless of region or climate, and financial investment? These are all significant considerations, but even more critical are the basic "nuts and bolts" skills necessary to make a working farm successful.

Basic knowledge of the entire farming enterprise is essential to creating, maintaining, and enjoying it as a business as well as a lifestyle. This knowledge has a wide range of applications.

Farmers have long been considered jacks-of-all-trades in a positive sense of the term. They need to be mechanics to maintain and repair equipment. They need to be agronomists to understand seeds, fertility, plantings, tilling, grass production, soil structures, and more.

They need knowledge of husbandry to raise animals humanely and efficiently. They need to be safety inspectors and instructors to avoid injury to themselves and their family members.

They need to be good bookkeepers to maintain a profitable business and have marketing skills to leverage their products into the greatest income while minimizing losses.

They need to be conservationists to protect their farms for the future generations, whether for their own descendants or for others.

And successful farmers need to be poets to appreciate the morning sunrise, the evening sunset, and to listen to their heartbeat upon the land they walk.

For sure this is not a small task. But it is not an impossible one either. Taking that first step—wanting to learn—opens a new vista to your future.

This book will help you understand the myriad issues involved in developing, operating, and enjoying a safe and successful farm that you may someday hand down to your son or daughter.

—Philip Hasheider

Chapter 1

The Business of Farming

STARTING A FARM BUSINESS

Farming is a lifestyle, but it is also a business to which you must commit time, energy, and financial resources. Without these three ingredients, it's difficult to be successful.

The most important thing to do on your farm is to make a plan. Know things like what kind of farming you want to do, how much money you'll need, what the market is, and how much equipment you have.

The first choice you'll need to make is what kind of farm you want to have. Farming offers enormous variety: it can involve raising livestock or grains, vegetables or fruit, flowers or woodlands, or other agricultural products. Farms can be huge, ranging in the hundreds or thousands of acres, or they can be as small as one acre.

You do not need a large acreage to farm with produce or livestock. However, having access to pasture areas will greatly enhance your ability to economically raise goats, sheep, pigs, or cattle.

The type of farming you pursue will largely determine the size you need. It generally takes more land and equipment to raise livestock than to grow vegetables. You need to provide feedstuffs such as grains and forages for live animals, while vegetables only need soil, water, and sunlight. Different types of production farming have different requirements. All these options mean you'll need a plan before you begin.

To develop a farm plan, you'll need to identify the type of farming practices you wish to pursue, project financial needs for those practices, research the viability of marketing strategies for your products, and figure out your prospective farm's production capabilities. Ideally, you'll do this before you plant a seed or buy a baby chick. (If you already have a farm, a plan can help make it more successful.)

For example, starting a cattle-raising business requires land, buildings, animals, and equipment. Vegetable farming requires equipment to till and prepare the soil and, depending on the extent of your production, perhaps additional labor, in which case you become a manager of people as well.

Learn about Farming

Good planning, research, and obtaining accurate advice will help you avoid unpleasant surprises. The more you read and learn, especially if you are new to farming, the better prepared you will be when you start. Although you can't know everything there is to know beforehand, you will have some basic knowledge. You will learn by trial and error as well, but this way of learning is as old as farming itself.

Advice can come from an agriculture lending group or bank, county agricultural extension office, or private professional services that specialize in farm production and setting up farming enterprises.

Writing a Business Plan

Any start-up business needs a business plan. A business plan is a way for you to state your goals; test ideas to determine if they are feasible; anticipate finances and markets; plan for crops, livestock or other enterprises; and build management. It lays out a pathway for you to achieve your goals.

A number of organizations can help you develop your farm business plan including university extension, usually administered through county offices, the United States Department of Agriculture (USDA), regional small farm institutes, and various private business planning companies.

Farm plans also include budget plans. Budget plans are essential to help you understand your production costs, what markets are available, pricing your products or what to anticipate when selling your products, cash flow, managing debt, bookkeeping, and many other things relating to your farm.

Funding

A farm development plan needs to be economically sustainable if you need to obtain funding to purchase land, equipment, and other resources from lending institutions, investors, or grant-making agencies. You might not have all the capital needed to start farming on the scale you would like.

Grants and loans for real estate purchases, production operation, and machinery or breeding stock purchasing are often available for beginning farmers through government agriculture agencies such as USDA through the Farm Service Agency (FSA) or through many local banks that service agriculture. Developing a sound business and farm plan will help you in any presentation you make to request a loan or grant.

Get Licenses and Permits

Local and state laws vary across the country regarding the establishment of small farm businesses. Generally the basics are the same and include having a business name, registering and/or purchasing a business license if you sell your products to the public, and securing an employer identification number. Also, purchasing insurance or product liability for your farm and business will be needed to protect your investment.

Zoning restrictions may be in place in the area you choose to farm, so you need to contact the county or township officials to determine if any restrictions apply.

If you want to raise livestock, be sure to talk to someone with experience, who can help you understand the details of the business.

Farms and Property Acquisition

Buying a farm is a large financial investment, so choose your farm carefully—one that is within your proposed budget—and get professional advice if you are unsure of how to proceed. When purchasing a farm, you may have a strong attraction to live in a certain area or region because of climate considerations, soil type, and the distance from a major city or village. In some cases, the appeal may be the distance away from population centers.

Renting a farm for a period of time may help you ease into farming, as well as help you position your finances toward an eventual purchase. By renting, you can determine the farm's productive capacity and workability of the facilities available, such as the house, barn, sheds, and fences, before committing to a purchase.

Many of the financial aspects of purchasing a farm are similar to the purchase of any other business, such as budgets required by the lending source and capital costs for initial purchases including fertilizers, seeds or animals, and equipment. Legal advice and assistance should be obtained before entering into any purchase commitments.

A number of factors need to be considered if you plan to purchase livestock. It's best to work with someone who is ethical and understands this side of the business.

Learn as much as you can about farming before you begin. Your chances of success will be greater if you understand how to take care of tractors and field equipment and follow safe operating procedures every time.

Location and Social Considerations

In many cases, your farm will also be your home. The property's location and the services available may be important factors. Living on a farm does not necessarily exclude you or your family from some of the conveniences or services available in an urban area. There just happens to be a greater distance to access them.

When purchasing a farm, assess whether the house or dwelling meets your family requirements both now and in the future. If you have a young family, it may be important to be close to schools, doctors, or transportation systems. The availability of professional veterinary services is important for your animals.

If community activities are important to you and your family, many areas have a chamber of commerce that provides information about activities during the year. Another consideration may be the opportunity for alternative or off-farm income or employment.

Physical Considerations and Application

Many considerations relating to the physical factors of buying a farm may be closely tied to your social considerations, such as location and climate. Farm size will determine things such as the number and type of livestock you can raise, the crops or vegetables grown, and other factors.

Soil type and fertility are important considerations because in some cases the soil type is tied to the value of the property. Soil type can also influence the types of crops raised and the durability of the crops during a drought or extended dry spell. Heavy soils help sustain crops in dry conditions while lighter, sandier soils do not.

The quality of the buildings and of improvements made on the farm for the buildings or land can be a determining factor in a farm purchase. Extensive building renovations may require finances that could otherwise be directed toward the operating expenses of your farm. Yet, the need for these same improvements may lower the purchase price and be an attractive option.

Whatever farm you purchase, it is necessary to fully understand the boundaries. Walking the fences will provide you with a better idea, such as if a grazing program may or may not be suitable.

Buildings and Equipment

The facilities needed for raising livestock such as cattle, sheep, goats, pigs, poultry, and other fowl do not need to be extensive, but they may determine to some extent the size of your program and the numbers with which you can easily work. Buildings do not need to be elaborate or expensive, but they do need to be sturdy enough to withstand the movement of livestock and the pressures their body types exert upon them.

Buildings that have seen little recent use may need to be remodeled or renovated. Buildings that are still structurally sound can be used as housing for animals or as storage and staging areas for produce.

Raising crops will require some equipment to help with the harvest. This applies whether you own or rent your farm. In some regions, the climate is moderate enough that a winter livestock feed supply is available and can be supplemented with stored feed. However, in most northern areas of the country, a winter feed supply must usually come from stored feed harvested during the growing season. To maintain a sufficient feed supply for your livestock, you will have to harvest crops during the growing season or purchase all of their nutrition needs.

If you decide to do the field work yourself, the list of equipment grows considerably. Tractors, tilling machinery, and repair tools are only some of the equipment required.

Sufficient experience in operating any equipment in a safe and efficient manner is a very important knowledge base for you as well as any other family member or employee. Farm machinery can be dangerous equipment to use if you don't have the knowledge or ability required to handle it properly. If you cannot find farm safety classes in your area, one alternative is to contract others for the field and crop work. Contracting dramatically reduces the amount of equipment you need to purchase, as well as your exposure to heavy machinery.

Remember that in northern climates, for the vast majority of the year, your equipment will sit idle. If you buy the equipment, you will be paying on an asset that is not working. Contract hiring of your crop work can help you get established without making the huge investment in machinery. Equipment expenditures can be substantial when getting started, and less money spent on machinery means more will be available to run your business. Generally the local county agricultural extension office can help with locating names of independent or custom operators in your area and provide typical custom rates.

Contract hiring for field work not only frees up capital, but also relieves you of any equipment repair costs. Generally custom hire businesses can accomplish more fieldwork in a shorter amount of time than you could on your own.

Renting equipment may be a more attractive option than purchasing. Renting allows you to use appropriate equipment for the crop and then return it without outright purchase. Check with your local equipment dealers for information about renting equipment.

Many small items needed on the farm can become an extensive list, including water tanks, hay feeders, forks, shovels, pails, wire, tools, gates, and halters for tie-ups. An easy way to get a better idea of what may be needed on your farm would be to attend a local farm auction and observe the many small items that are sold at the beginning of the sale prior to the larger equipment such as machinery and tractors. You may not need all of the items you see being sold, but it doesn't cost you anything to attend and get ideas.

PRODUCTION SYSTEMS: ORGANIC, SUSTAINABLE, AND CONVENTIONAL

Most agriculture will fall into these three categories: organic, sustainable, and conventional. The organic and sustainable systems are gaining popularity with farmers and others for several reasons. The most prevalent are increased markets for products from these systems and the increased value those products bring to the marketplace. Significant differences are apparent among organic, sustainable, and conventional farming practices.

Conventional farms have used intensive planting and harvesting of crops, aided by crop chemicals and commercial fertilizers, as a way to increase production and generate higher profits. Much of this has been driven by consumer preference for cheap food. Shifts in consumer tastes and the higher demand for foods coming from organic and sustainable farms are making these production systems attractive to farmers. This consumer shift, which cultivates the image of healthier foods, has brought more money to organic and sustainable markets. It has allowed farmers to utilize practices that may be more in tune with their personal ethics and nature's harmony and also be a better fit for small-scale farming.

One of the most attractive aspects about all three farming systems is that you can stop one and switch to the other system; although it is easier to switch to a conventional farming system from one that is organic or sustainable than the other way around. This ability to switch may be important if a change is required in your farm sometime in the future. You have the option to alter your farming plan to adjust to the new circumstances.

Organic farm production is increasing as the demand rises for produce, meat, eggs, and fruits raised in a chemical-free environment. Organic highbush blueberries (here, bushes being planted in spring) are an excellent market crop.

Organic Farming

Organic farming is an ecologically balanced approach to farming that supports, encourages, and sustains the natural processes of the soil and the animals. Those involved with organic farming have a perspective that reaches beyond the present and requires a long-term commitment to their program.

In many ways, organic farming has been a reaction to the industrialized and large-scale, chemical-based farm practices that have become the norm in food production since the end of World War II. Organic farming emphasizes management practices over volume practices, considers regional conditions that require systems to be adapted locally, and supports principles such as environmental stewardship.

Farming organically is a method where the whole ecosystem of the farm is incorporated into the production of animals or crops. It seeks to obtain the greatest contributions from on-farm resources such as animal manures, composts, and green manures for soil fertility and to eliminate external additives, especially synthetic chemicals.

Organic farming is flexible enough that if sufficient quantities of materials cannot be produced on the farm, off-farm nutrients, such as natural fertilizers, mineral powders, fortified composts, and plant meals from approved organic sources can be applied without risking certification.

On an organic farm, you enrich the soil with organic matter and compost made from such material as manure, leaves, rotted produce, and grass clippings.

Other strategies used in organic farming are cover crops and crop rotation. This field of winter wheat will be harvested in spring; the field will then be planted with soybeans.

Organic farming promotes the health of the soil by encouraging a diversity of microbes and bacterial activity. In turn, this diversity enhances the growth of plants and grasses deemed healthier for the grazing animal.

The exclusion of insect and pest control products, as well as antibiotics for use on animals, can make raising livestock, green produce, or other products seem more of a challenge. But it is being done in many different parts of the country with great success because of a change in traditional thinking by those working with that type of farming. Opportunities for organic farming are available, and it is possible to accomplish your goal of organic farming by studying what you put onto your land and what meat or milk you harvest from it for the marketplace.

Organic Certification Process

The term *organic* is defined by law in the United States, and the commercial use of organic terminology is regulated by the government. Certification is required for producing an organic product that is packaged and labeled as such for sale. The process to become certified can seem extensive, but the result is that products produced under certification are authenticated.

Standards for organic certification are set by the government or various organizations in which farmers can become members. Lists of these organizations are available from most county agriculture extension offices or from your state's department of agriculture.

If you choose to farm organically, you need to have your soil analyzed to determine which nutrients need to be added to balance its components for optimum plant growth and nutrition. You will be feeding the soil so that it can feed the plants. Grains may be fed to livestock but, under organic rules, only from sources that produce it organically in order to maintain your certification. Specialized organic markets must be found to purchase feedstuffs used on your farm and for the sale of livestock to get the greatest benefit from your efforts.

If your farm has used conventional practices with chemicals, you'll have a three-year transition period before organic certification can be obtained. During this time no chemicals of any kind can be used on the land or on the animals, antibiotics cannot be used to treat the cattle, and genetically modified organisms (GMOs) cannot be planted.

Sustainable Farming

Sustainable farming is a goal rather than a specific production system. Although they are often thought to be the same thing, sustainable and organic farming are not necessarily synonymous. The goal of sustainable farming is to approach a balance between what is taken out of the soil and what is returned to it without relying on outside products. The aim is for perpetual production.

The idea of sustainable farming starts with the individual farm and spreads to communities affected by this farming system, both locally and globally. Many people concerned with sustainable issues take a global approach with this system because it does little good for the overall goal if your farm has a negative impact on the environmental quality somewhere else. Sustainable farming requires that outside products put onto your land are available indefinitely and nonrenewable resources need to be avoided.

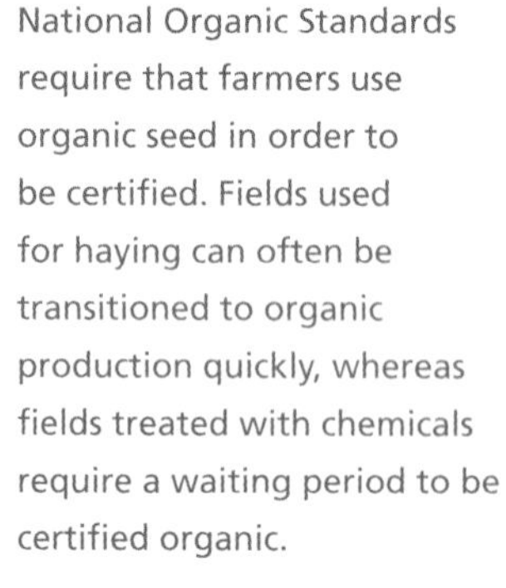

National Organic Standards require that farmers use organic seed in order to be certified. Fields used for haying can often be transitioned to organic production quickly, whereas fields treated with chemicals require a waiting period to be certified organic.

Honeybees are one of the most beneficial insects on any farm, and can be killed by the same pesticides that kill other insects. On an organic or sustainable farm, bees are more likely to thrive—and help you in turn.

Growing and harvesting crops removes nutrients from the soil and without replenishment the land becomes less fertile. Therefore, one of the major keys to success in sustainable farming is soil management. Those following this practice believe that a healthy soil is a key component of sustainability. A healthy soil will produce a healthy crop that has the optimum vigor and is less susceptible to insects and pests.

If you are buying a farm or if you are already farming, it is likely you will need a transition period to become sustainable. This transition is a process that generally requires a series of small, realistic steps. Because of some costs involved, your family economics and personal goals may influence how fast or how far you will go in this transition.

Raising livestock using sustainable production principles will include consideration of how you integrate your crop and animal systems. The key components are profitability, management of your animals, and stewardship of the natural resources, such as water and soil.

Proper grazing methods and pasture management can eliminate the most adverse environmental consequences associated with your animals, including stocking rates, so that limited amounts of feed must be brought in from the outside. The long-term carrying capacity of your farm must take into account short- and long-term dry spells and reduce overgrazing on fragile areas of the farm.

Sunflowers attract pollinators and are also excellent cutting flowers. Raising nectar-producing plants and flowers will help pollinator colonies sustain themselves.

Animal health is an important part of a sustainable farm because the health of your animals greatly affects reproductive performance and weight gains of calves. Unhealthy livestock wastes feed and requires additional labor.

Sustainable farming asserts that the stewardship of natural and human resources are of prime importance. Stewardship of people includes consideration of social responsibilities, such as working and living conditions of farm laborers, the needs of the rural communities, and consumer health and safety both now and in the future.

Sustainable agriculture requires a commitment to changing public policies and social values, as well as preserving natural resources. This may include becoming involved with issues, such as food and agricultural policy, land usage, labor conditions, and the development and resurgence of rural communities.

Developing a strong consumer market for your products can help sustain your farm. Becoming part of a coalition can be useful in promoting your products and educating the public in general. By educating more of the public, you will likely increase your market and therefore continue the cycle you started.

Conventional Farming

Many conventional farming practices are geared for maximum production by using large amounts of outside resources to produce products. Whether it is corn, soybeans, alfalfa, wheat, or any number of other crops, intensive farming practices have become the norm for many farms.

These farms typically use chemical-based products to control insects, pests, and weeds and to promote rapid growth of crops and animals or increased milk production. Hybrid seeds or GMOs are sometimes planted to increase yields. Highly specialized machinery does most of the work, and the operator's feet may seldom touch the ground.

Conventional farming differs dramatically from organic and sustainable farming in that it requires a heavy reliance on nonrenewable resources—such as fertilizers, gas, and diesel fuels—and practices such as excessive tilling, which can lead to soil erosion. There is some evidence that these practices cause long-term damage to the soil.

MARKETING YOUR FARM'S PRODUCTS

You can market your farm's products in several ways depending on what you produce. Meat can be marketed from your farm, through a cooperative, or directly to consumers. Milk can be marketed through private companies or cooperatives. Produce such as vegetables or flowers can be sold at farmers' markets or through a community supported agriculture

Quality and marketing are the most important ingredients to your success. Good marketing gets your name out and spreads the word that you have a quality product for sale. Here, a busy farmer fills orders for his tomatoes.

Developing a niche market for your products can potentially provide extra income for your farm. Products can be sold privately or publicly at large or small farmers' markets.

(CSA) program. Agritourism may bring large numbers of people at specific times of the year. On-farm stores and pick-your-own fruits and berries may also draw customers to your farm. Each of these venues has opportunities and challenges as well as regulations.

Choosing the time and method of marketing your farm products can increase your profits. For example, too often livestock growers sell their animals at the most convenient time rather than the most profitable time. To become a good marketer, you will benefit from understanding the marketing system and how your prices are determined and then reviewing the options available and which to pursue. Bear in mind that consumers eventually have the most impact on market prices because of their tastes, attitudes, and sense of their family's welfare.

Value-added products are terms used to identify those grown or raised on a farm and are worth more than the regular market price because of an added feature or practice. Value can be added through a physical processing change or some characteristic that has value to your customer, such as an animal or produce being raised to a specific standard.

A farmers' market is an excellent place to showcase your farm's produce. Use your farm's production protocols to explain why your products are unique. This venture can be part of your value-added marketing strategy.

Two keys to a successful market day are your smile and your ability to remember names. Being upbeat, answering questions, and treating people respectfully are the best ways of attracting and maintaining customers.

Adding value to your products is one way to increase profits. This perceived value is tied to increased quality and desirability by the customer. This option may work for you because by offering a value-added product, you can bypass traditional routes and capture more of the profits for yourself.

Value-added selling may require some marketing skills on your part and finding a niche market where you can sell your products. Finding your own customers generally requires more effort than traditional marketing systems, but the rewards can be greater.

Some salesmanship may be required to explain the advantages of your produce or meat to potential customers. Why is it better? It may be because you raise your livestock using a grass-based program or they are pasture-grazed to enhance the eating quality of the meat. You may be able to provide products not normally in season. You may use other concepts that capture the attention and imagination of potential customers, such as certified organic. Value is created when a product meets or exceeds the customer's expectations.

Niche markets for such products as natural, organic, green, or pasture-grazed are becoming recognized by customers for the healthy conditions under which the animals or produce are grown. Marketing groups may be available to help you sell your product. Most county agricultural extension offices can provide you with contact information for groups in your area.

Farmers' Markets

Direct marketing to consumers can be as simple as selling at a farmers' market. This is a good marketing entry point because you set your own price and sell what you have available. In addition, a lot of social interaction occurs between farmers and customers at these venues, allowing both to learn from each other. The farmer learns what the customers are looking for, and the consumers learn about your farm, what you produce, and the quality you bring to the market. Many communities have started a farmers' market and you should check what is available in your area.

Every marketing strategy has challenges, and these types of markets do too. Your product may not sell out, or the customer's loyalty is to the market and not necessarily to you. If selling at a farmers' market, you or another person will need to be present regardless of weather conditions at the time.

Most farmers' market venues have rules and regulations that you must follow in order to participate. Check with the market manager about their rules and state or local regulations that may apply.

A summer CSA box can be filled with a variety of different vegetables, fruits, cheeses, or whatever your farm produces. Offering CSA shares is an effective way to stabilize your finances during the planting and growing seasons.

Community Supported Agriculture (CSA)

A CSA is a system where the producer supplies consumers with produce or other farm products weekly at a predetermined site. Local households and farmers work together toward a common goal of sharing. Farmers are paid an annual fee by each household in the winter or spring, which entitles that household a share in that farm's seasonal harvest. Weekly deliveries are made at prearranged sites or held for pickup at the farm. Some farmers allow their customers to visit the farm and to participate in the harvest under their supervision.

A well-managed CSA involves four key ingredients: product growing expertise, customer service, planning, and recordkeeping. You need to know how to grow the products you offer for sale in sufficient volume to satisfy your customers' needs as well as having a diverse variety due to differing growing patterns of your produce or livestock. Customers join a CSA for the quality food they receive and for that real connection with a farm where their food is grown. This powerful attraction is a major selling point and can lead to many years of continued support.

If you are developing a CSA on your farm, whether with produce, meat, eggs, flowers, or homemade cheese, you must plan an entire season's production before you begin to plant or one animal is purchased. Understanding the growing patterns of your vegetables and varieties and their rates of maturity will help you develop a system of timing and succession plantings to ensure you have a consistent harvest throughout the season.

You need to keep detailed records of your production and finances to succeed with a CSA. It is a business, and customers are investing in your program because they expect a return on their money in the form of farm products. You are looking for a profit for your time and effort. These records may include the names of your customers, amounts paid, whether full or half shares, delivery points, estimated cost of production for each variety or animal, estimated salary you expect, and many other factors. Budgeting projections are necessary so that you have enough money to last through the season and to cover situations beyond your control, like dry spells or extensive rain.

A typical CSA vegetable farm can serve about twenty to thirty households (shares) per acre in production. These acres will need to be planted with ten to twenty types of crops to ensure a sufficient and variable production. If you offer meat products, one beef animal, depending on size, can serve thirty or forty households per month. In this case, you will need to develop a cold storage transfer to keep the meat from spoiling, and the animal will have to be processed at a state- or USDA-inspected facility.

Agritourism

Agritourism is a growing segment of agricultural production. In this case, you are providing an emotional experience for those customers who do not

Agritourism allows people to come to your farm to experience a rural setting. Hayrides, cook-outs, and corn mazes are just some of the events that you can offer.

have access to farms. People come to your farm and interact with it through a variety of activities such as animal feedings and petting zoos, hay/sleigh rides, corn mazes, food and harvest festivals, floral arranging workshops, you-pick opportunities, and a wide variety of other activities.

An agritourism farm can be a destination and offer activities to entertain, educate, and enlighten your customers. Developing an agritourism farm will require assessing local rules and regulations for public gatherings as well as your risks in hosting a number of people on your farm. You will need to offer safe, fun activities and provide a low-risk environment. You will also need to plan your business with your insurance provider.

Intermediate Marketing

Selling your farm products to a specific buyer for resale is called intermediate marketing. If your production is large enough, you may be able to provide a steady stream of product to restaurants, grocery stores, and other markets without having to use a farmers' market or CSA. However, intermediate marketing typically includes both of those venues in a marketing plan.

Private labels are required for producers selling meat, jams, jellies, or any other product that requires processing. A distinctive and attractive label may help increase sales.

Market Your Own Label

Capturing a niche market can have financial and personal rewards, especially if you develop your own label. Private labels are becoming more prevalent, as producers with small numbers of animals as well as berry farmers, beekeepers, condiment makers, and others begin to sell their products.

If you have a small farm operation, you can invest more individual care and oversight into each product—and you can turn your personal attentioninto a solid marketing tool by identifying your farming practices.

If you plan to sell and deliver products to customers who live some distance from your farm, you might need to think about a storage container that can maintain cool or freezing temperatures. Contact your state department of agriculture for details on how to start direct marketing. They can advise you on all regulations, licensing, and labeling requirements.

If you're raising animals for meat, you will need access to a slaughter facility that can guarantee your animal's meat will be separated from others processed the same day. They can explain the costs involved in processing, packaging, and stamping your meat with your label for proper identification. Be sure that the facility is state or federal inspected. Developing your label and storage facilities should be done well in advance of slaughtering any animals.

No matter what product you have, you'll want to advertise, whether by word of mouth, print ads, or some other means. Don't wait until your product is ready to sell to start advertising. Plan months ahead. Be sure to highlight the emphasis you place on your production system and that the consumers understand its importance.

Clean, dry wool from your own sheep doesn't have to be labeled, since it isn't prepared food, but a label is one way to set your goods apart.

Rules and Regulations

Food safety is of primary importance when marketing your product. You need to know and understand a variety of local, state, and federal rules and regulations for selling farm products to the public. Some are simple, while others are more detailed and specific. Some cover how food products are handled from harvest to sale. Other regulations cover storage and processing. The best way to know what applies to your situation is to contact your state's department of agriculture and request the current rules and regulations for farmers wanting to market to consumers, grocery stores, restaurants, and other public venues.

Chapter 2

Crops

In this chapter, we'll introduce you to a variety of crops that you can grow on your farm, and give you the basics on compost, fertilizer, planting requirements, and more. Whether you're interested in growing grain on a small scale or wish to raise vegetables for the market, the information in this chapter will give you a useful foundation.

A field of barley ready for harvesting.

GRAINS

When you think of the word "crop," you probably envision a vast field of golden wheat or a silo filled with freshly harvested grain. You might think that you can't grow grain unless you have a huge farm, but that's not the case: growing grain on a small scale can be an intensely rewarding experience.

Grain is typically divided up into two categories: cereal grains and pseudocereal grains.

The Cereal Grains

For many people, the most important cereal grain is **wheat**, thanks to its status as the backbone of bread, baked goods, and pasta. Wheat is found in several types, including hard, soft, red, white, spring, and winter. Ancient types of wheat include einkorn and emmer, along with durum, which is useful for making pasta and can be used for bread if it is combined with another type of wheat flour that contains higher quantities of gluten. Some gluten-intolerant individuals have reported that they are able to tolerate ancient varieties of wheat, such as einkorn and emmer, better than they can tolerant modern-day varieties of wheat.

Rye is a smart choice when you're faced with a short growing season, which might explain rye's popularity in Europe. In North America, rye is generally grown for livestock, but it's also used to make rye flour, which has a lower gluten content than wheat. This hardy grain grows quickly and doesn't require hulling.

There are three main types of **oats**: white, red, and hull-less. By choosing a hull-less variety, you can save the hulling step in the harvesting process.

Grains: seven grain mix (bottom left); hard red winter wheat (bottom right); pearled barley (top).

Grains: millet (front); quinoa (center); and buckwheat groats (back).

Oats are relatively easy to grow, and if you harvest carefully, you can ensure that your oats are gluten-free.

Barley is an adaptable grain that grows in a wide range of conditions. It hasn't achieved the widespread popularity of wheat, but it's used in soups, casseroles, and alcoholic beverages. A hull-less variety of barley has recently been introduced.

Millet has long been important in Asian and African cuisine. It prefers a warm climate. Millet can be found in many varieties: the most common include pearl millet, finger millet, foxtail millet, and proso millet. Millet is gluten free, but if you want to make bread with millet, you'll want to stick with flatbreads.

The Pseudocereal Grains

After centuries of popularity in South America, **quinoa** is finally gaining momentum in North America. Quinoa (pronounced KEEN-wah or key-NO-uh) received official United Nations designation, making 2013 the International Year of Quinoa (the first time since 2008 that a specific food has been granted such an honor).

Although not nearly as popular as quinoa, two other pseudocereal grains merit a mention: **amaranth** and **buckwheat**. The former is a highly nutritious, protein-packed, gluten-free grain with high quantities of lysine, and the latter—despite its name—is not related to wheat at all. Buckwheat is familiarly used for pancake flour, but it also makes one of the best cover crops. Amaranth can also be ground into flour for baking, or you can try cooking the grain and popping it (similar to popcorn). You must cook amaranth before eating it.

Getting Started with Grain

If you wish to harvest grain in significant amounts, you'll need a lot of space. If you'd like to grow grain to sell it, you'll need to harvest a minimum of three to five acres. (See chapter 6 for more on harvesting grains.) For home use, you can plant less, of course. How much you'll harvest per acre depends greatly on the grain, where you live, how good your soil is, and if you have irrigation or plentiful rainfall. From a nonirrigated acre of reasonably good soil, you might expect a harvest of thirty to fifty bushels of wheat (60 pounds in a bushel).

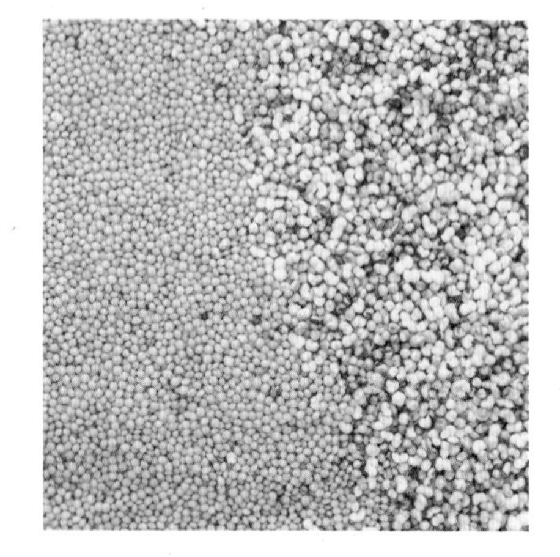
Amaranth: raw (left) and popped (right).

Before you plant, have your soil tested so that you can accurately determine any soil amendments that need to be made for the crop you're going to grow. You'll also want to consider seasonal conditions. Millet, for example, thrives in warm soil and should typically not be planted until the soil has reached 60 to 65°F. Rye, on the other hand, likes cool weather and cool soil temperatures.

You'll generally get better results if you plant grain with a grain drill or grain planter, but you can use broadcast seeding in some circumstances. If you're planting only a small area of grain, you can just sow seed by hand.

VEGETABLES

There's something inherently satisfying about growing your own vegetables. Perhaps it's the fresh-from-the-garden taste, or maybe it's the knowledge that you've grown food—*food*!—from a tiny seed. Or maybe you're anticipating selling your homegrown vegetables at farmers' markets or through a CSA program. Whether you're raising vegetables for your home or for the market, it's a fun and rewarding venture.

Here are some general terms and definitions you'll need to know.

- **Annual:** A plant that only survives for one growing season and does not overwinter
- **Bolting:** Going to seed
- **Clay soil:** Thick and heavy and retains (too much) water; the opposite of sandy soil
- **Compost:** Organic material used for planting; adds much-needed nutrients to the soil and consists of items such as dried leaves, hay, straw, grass clippings, eggshells, coffee grounds, vegetable peelings, as well as manure
- **Cold frame:** Like a miniature greenhouse; protects plants from early and late frosts
- **Crop rotation:** The deliberate alternation of plant locations; used to minimize soil depletion and minimize the risk of disease
- **Direct-sowing, direct seeding:** Planting seeds directly into the soil rather than starting them indoors first

The owners of this small farm sell their vegetables at the market and through a CSA program.

Heirloom tomatoes are typically less resistant to disease, so be sure to rotate your crops every spring (on each section grow plants from a different vegetable family than the previous year).

- **Double-cropping:** Planting another vegetable after the first one is harvested
- **Growing season:** The length of time between frosts; the time between the last frost of spring until the first frost of fall
- **Heirloom:** Generally open pollinated varieties, meaning that their seeds breed true; tends to be older varieties that have been handed down from generation to generation
- **Hills:** A raised heap of soil, approximately 12 to 24 inches in diameter and 4 to 8 inches in height
- **Hybrid:** The genetic crossing of two distinctly different plants, resulting in a plant with increased hardiness and disease resistance but lacking in the ability to reproduce itself
- **Intercropping:** Planting more than one crop in a particular area
- **Loam soil:** A nice balance between sand and clay; suitable for many types of plants and is an ideal choice
- **Mulch**: A light layer of straw, hay, wood shavings, or grass clippings to keep weeds at bay and help the ground to retain moisture
- **Perennial:** A plant that survives for more than two growing seasons
- **Raised bed:** A plot elevated somewhat higher than the ground around it to create a space about 6 or 8 inches above the surrounding area
- **Sandy soil:** Drains well, which can be helpful, but also has a tendency to dry out easily
- **Succession planting:** Planting seeds continually throughout the season to ensure a continual crop

VEGETABLE FAMILIES

FAMILY NAME	COMMON VEGETABLES
Alliaceae	Onions, Garlic, Chives
Apiaceae/Umbelliferae/ Parsley Family	Parsley, Carrots, Dill, Parsnip
Chenopodium/ Goosefoot Family	Beets, Spinach
Brassicaceae Family	Broccoli, Brussels Sprouts, Cabbage, Cauliflower, Radishes
Cucurbitaceae/ Cucurbit Family	Cucumbers, Squash, Pumpkins
Fabaceae/ Leguminosae Family	Peas, Beans
Poaceae/Gramineae/ Grasses Family	Corn
Solanaceae/ Nightshade Family	Potatoes, Tomatoes, Eggplant, Peppers

- **Thinning:** The process of removing some young seedlings to allow ample space for the strongest seedlings to grow and thrive
- **Tilling:** The process of loosening the soil by working it and making it soft and supple; can be accomplished by hand, using tools such as a shovel and pitchfork, or by machine, such as a rototiller
- **Trellising:** Providing support for growing plants in the form of poles, string, fencing, ladders, or cages
- **Trowel:** A small, handheld shovel

Proper spacing of rows and the use of soaker hoses can help prevent plant disease.

Green beans.

Meet the Veggies

Ready to get growing? Here is at-your-fingertips information for seventeen common vegetables.

Beans

Types: Pole and bush beans are the two main classifications; within these two groups are numerous varieties. 'Empress' is an excellent bush variety, noted for its abundant production and incredible taste. A fun pole variety is 'Lazy Housewife'. Unfortunately it tends to produce exactly as its name implies: lazily. For a bit of color, plant 'Royalty Purple Pod', and if you want an exciting choice for dried beans, try 'Calypso' or 'Hidatsa Shield Figure'.

Requirements: Beans prefer to be planted in full sun and perform best in well-drained soil.

Special Considerations: Pole beans need the support of a trellis, fence, or poles to climb. Bush beans do not require these supports; they produce more abundantly but for a shorter time. Pole beans produce for a longer period of time, especially if you continually harvest the beans.

Planting Information: Beans should be planted well after the danger of frost is past. Do not plant beans too early; frost is very dangerous to beans. Bush beans are somewhat hardier than pole beans, so you can safely plant bush beans after the average last frost date has passed; wait another two weeks or so until the latest possible frost date has passed before planting pole beans. Plant your seeds 1 to 1½ inches deep, and place two seeds per hole. Make your holes 2 to 4 inches apart, and space your rows 2 to 3 feet apart.

Growing Time: 50 to 90 days, depending on the variety. Bush beans are the quickest; pole, lima, and soy beans require more time. For your bush beans, if you wish to eat them as fresh snap beans, the maturity date is faster. If you wish to save them as dried shelled beans, add another 30 days.

Soil pH: Ideally, 6.0 to 6.5 (slightly acidic).

Cabbage.

Cabbage

Types: Several types of cabbage include early, mid-season, and late varieties. Their sizes correspond to their names: early maturing varieties of cabbage usually weigh about 4 to 6 pounds; later maturing varieties can reach up to 16 pounds. 'Stonehead' hybrid is an early maturing, green 5-pound cabbage. 'Mammoth Rock Red' is a 7-pound heirloom variety.

Requirements: Cabbage is best planted in a sunny area, although some shade is tolerable. Aim for good, fertile soil with lots of nutrients.

Special Considerations: Watch the pH of your soil, and add some lime if your soil is too acidic (under 6.5 pH) for cabbage to be happy.

Planting Information: It's best to start seeds indoors rather than sowing seed directly. Plant seeds ¼ to ½ inch deep and 3 inches apart. Later, transplant the seedlings, planting 18 inches apart for early varieties and 24 inches apart for late varieties. Allow 24 inches between rows.

Growing Time: 58 to 73 days, depending on variety. Some super-early varieties mature in 40 days.

Soil pH: Over 6.5, up to 8.0.

Carrots.

Carrots

Types: From tiny 2-inch 'Thumbelina' carrots all the way up to large 12-inch varieties such as the 'Envy' hybrid, carrots can be found in a wide range of lengths, shapes, and colors including purple, white, red, yellow, and the traditional orange. For unsurpassed flavor and a fabulous color, try 'Dragon' heirloom carrots!

Requirements: Carrots prefer a cool and wet environment coupled with a light, sandy, fine, fluffy soil. The soil must be free of rocks, clay, or manure that hasn't been composted thoroughly.

Special Considerations: About three weeks before your average last date of frost, sow your carrot seed directly into the soil. Mulching will help to retain the moisture in the ground. Thinning is almost a necessity.

Planting Information: Carrot seeds are teensy tiny, and although you'll ideally plant them about ¼ inch deep and about ½ inch apart (in rows about 12 inches apart), it's not always easy to space them exactly as you would like to. Germination takes about two weeks, so be patient.

Growing Time: 58 to 75 days.

Soil pH: Ideally, between 6.5 and 6.8 (slightly acidic), but slightly over or under this range is acceptable.

Cauliflower

Types: Cauliflower can be found in varying colors, including green, white, or purple heads. The 'Cheddar' hybrid is a bright orange-yellow color and very impressive!

Requirements: Cool and damp are the main requirements for succeeding with cauliflower. Choose a portion of your farm that receives partial shade, and then work at keeping the ground moist; mulch can be helpful in this respect. Ample water is important for growing cauliflower.

Special Considerations: As a member of the cabbage family, the cauliflower is somewhat hardy and can withstand minor frost.

Cauliflower.

Planting Information: As with cabbage and broccoli, start your cauliflower seeds indoors, ¼ to ½ inch deep, 3 inches apart, and then transplant outdoors in rows 18 to 24 inches apart.

Growing Time: This varies widely, from 55 days all the way up to 100 days.

Soil pH: 6.0 to 7.0 (slightly acidic to neutral).

Corn

Types: Sweet corn is easily the most popular (and delicious!) type of corn for culinary purposes; popcorn is not considered an "eating" corn, except as popcorn; Indian corn has multicolored kernels and is generally used for decorative purposes.

GROWING CORN ON A LARGER SCALE

It's simple to grow a few rows of corn on the edge of your garden or to implement the Three Sisters method when growing corn along with squash and pole beans, but maybe you'd like to try harvesting corn in larger quantities. Maybe you'd like to grow corn for feed, or maybe you're trying to avoid genetically modified (GMO) corn in your diet and you'd like to grow non-GMO corn for your own consumption. Or maybe you'd like to help meet the increased demand for corn by planting larger quantities to sell. In any case, expanding your corn crop isn't as difficult as you might think.

Of course, being a fan of heirloom vegetables and organic growing practices, I generally recommend open-pollinated varieties. Heirloom seed companies offer a wide range of options for sweet corn, popcorn, and Indian corn varieties, but bear in mind that sometimes these older varieties do not provide the uniform quality and yield that has become a bragging point for developers of newer hybrid varieties. So if high production and uniformity are your goals, then you might want to consider one of the non-GMO hybrid corn varieties, which can provide your desired yield.

Planting corn on a somewhat larger scale can require additional equipment. At the very least, you'll probably want a one-row garden planter to aid with the increased demands of planting. And while it's certainly possible to harvest corn by hand—farmers have long harvested corn in this manner!—it's undeniably easier to harvest using a corn picker or a combine. Admittedly, there is increased expense involved with machinery purchases.

Requirements: In order to thrive, corn needs sunshine. A windbreak is also helpful, as strong winds can blow down your stalks and damage them. Sandy soil is better than clay.

Special Considerations: Heirloom corn is also known as "open pollinated" corn; in other words, the seeds are savable. Hybrid corn is not self-sustaining, although it produces at a significantly higher rate than heirloom corn.

Planting Information: After last frost, sow in blocks, 1 to 2 inches deep, approximately 4 inches apart. Place four seeds per hole, and space 24 inches apart. The blocks are essential because corn is wind pollinated.

Growing Time: 65 to 90 days, but some super-early varieties may be ready in less than 55 days. Early maturing hybrids are on the shorter end; open pollinated varieties are on the longer end.

Soil pH: 6.0 to 7.0 (slightly acidic to neutral). Nitrogen is vital.

Corn.

Cucumbers

Types: Pickling and slicing varieties are the two most common types of cucumber. Pickling cucumbers are typically smaller; slicing cucumbers are larger. Additionally, you can plant bush varieties and vine varieties; the latter can be trained to climb a trellis, which is helpful for keeping your cucumbers out of the dirt. If you want to try the most delicious cucumbers in the entire world, opt for 'True Lemon' or 'Boothby's Blonde'—truly tasty! Bushy is a prolific heirloom variety but its flavor leans toward being a bit more bitter.

Requirements: Cucumbers love warm weather and lots of sunshine. Well-drained soil is important, as is ample water. If you harvest your cucumbers regularly, you'll keep them producing longer. Also be careful to protect them from frost. Cucumbers do not tolerate frost.

Special Considerations: You can start your cucumbers indoors, but you must be careful not to disturb the roots at transplanting time as this can stunt the growth of the plants.

Planting Information: Plant your seeds in hills, about 1 inch deep, with approximately six seeds per hill. Space your hills in rows approximately 3 feet apart.

Growing Time: You can find early maturing hybrids that can finish out in as little as 45 days, but most cucumbers are in the 50- to 65-day range.

Soil pH: 5.5 to 6.5 (moderately acidic).

Cucumbers.

Kale

Types: Three different types of kale are available: a headless variety, a rutabaga type, and collards. 'Blue Curled Scotch' is known for being hardy and easy to grow. Lacinato kale (also known as Tuscan kale, dinosaur kale, cavolo nero, or black kale) is also very popular.

Requirements: Kale is super hardy and will grow virtually anywhere. It likes sun but tolerates the shade and prefers well-drained soil.

Special Considerations: Since kale prefers cooler weather to hot, it's usually best when grown in the spring, early summer, or late autumn. The flavor is better when the temperatures are relatively cool.

Planting Information: Sow seeds ½ inch deep, approximately 1 foot apart, in rows 24 to 30 inches apart.

Growing Time: 55 to 70 days.

Soil pH: Over 5.5.

Kohlrabi

Types: Kohlrabi is available in several varieties: white, purple, and king-sized. Some say that kohlrabi tastes like turnips; others say it tastes more like apples.

Requirements: Kohlrabi prefers good sun and a soil that is well-drained. It tolerates cold weather very well.

Special Considerations: Many recommend that you plant kohlrabi in succession, due to the fact that the plants produce for only a short window of time. If you continue to plant over a period of several weeks, you'll be able to keep eating kohlrabi for the entire summer.

ABOVE: White kohlrabi.
RIGHT: Kale.

Planting Information: Kohlrabi can be started indoors or out. It can even be started outdoors weeks before your last date of frost. Sow seeds ¼ to ½ inch deep, with 3 to 6 inches between seeds and 12 inches between rows.

Growing Time: 38 to 62 days. Giant varieties can require upward of 100 days.

Soil pH: 6.0 to 7.0 (slightly acidic to neutral).

Lettuce

Types: Leaf lettuce varieties are celtuce, looseleaf, and romaine; head lettuce varieties are butterhead and crisphead. (Crisphead, also known as iceberg, is the type that you see in the grocery store: the big head of pale green lettuce with thick, tasteless leaves.) 'Black Seeded Simpson' leaf lettuce is a tried-and-true variety (pick it early, as it becomes slightly bitter if left too long). I also love 'Forellenschluss'; unfortunately, so do the squirrels!

Requirements: Lettuce doesn't need (or want!) good sun; instead, find a shady place. You can nestle your lettuce plants between other plants to save space and to give them some shade. Loose soil with plenty of nitrogen is best.

Special Considerations: Heat is not a friend to lettuce. It ruins the harvest by causing it to bolt. Iceberg lettuce is particularly prone to bolting, and many farmers find that it is much easier to grow looseleaf lettuce instead, which is much less prone to bolting.

Planting Information: Plant early, prior to the last date of frost, and continue to plant regularly over the course of several weeks to keep nice quantities of lettuce coming up all summer long. Sow ½ inch deep, ½ inch apart, unless planting leaf lettuce, which can be sown in a continuous row (although it will have to be thinned later).

Lettuce.

Growing Time: Leaf lettuce: 40 to 45 days. Head lettuce: 80 to 95 days.

Soil pH: 6.0 to 7.0 (slightly acidic to neutral).

Onions

Types: Long day or short day—what's the difference? Contrary to what you might think, long day onions are the type that you'll want to grow if you live in a northern climate, and short day onions are the type that you'll want to grow if you live in a southern climate. There are also day-neutral varieties, such as 'Candy' (our favorite!) that are suitable for growing in a wide range of locations. You can find red, yellow, or white onions in both long day and short day varieties, so just shop around until you find the combination that you're looking for.

Requirements: Sandy soil or a sandy loam is best for growing onions. Top this off with ample sunshine and regular watering (mulching with grass clippings helps to keep the soil moist), and you'll be well on your way to growing some marvelous onions.

Special Considerations: For those with short growing seasons, choose to start with onion sets or plants rather than seed. Onion sets are tiny bulbs that you can purchase at garden centers or in seed catalogs; onion plants are already growing and can save you a bit of time.

Planting Information: If planting from sets or plants, space the individual plants 2 to 4 inches apart, in rows 12 to 18 inches apart. Onions tolerate cold, so you can plant these sets before your

Onions.

Peas.

last date of frost. If you wish to start from seed, then you'll want to start them indoors a few weeks earlier.

Growing Time: Pay attention to the wide variance: 85 to 160 days, depending on the variety.

Soil pH: Above 6.0 up to about 7.0 (slightly acidic to neutral). Supplement with lime if necessary.

Peas

Types: You'll find many different types of peas to choose from—dwarf varieties that are bushlike as well as taller varieties that require support. 'Green Arrow' peas, for instance, only need minimal support as the plants do not grow beyond approximately 24 inches. If you want to plant peas with edible pods, look for "sugar" or "snow" peas. English peas, also known as "garden" or "shell" peas, must have the pods removed before you eat them. I have enjoyed planting Golden Sweet heirloom peas; for the best flavor, they should be harvested before the pods reach full maturity.

Requirements: You can plant peas very early in the spring; sow them outdoors six to eight weeks before the last date of frost. Peas enjoy cool weather and plenty of water. Your early spring crops should be planted in full sun, but if you try for crops in the summer, you might want to plant them in a location with a bit of shade.

Special Considerations: Dwarf varieties may not need support, but most varieties will require the presence of a trellis or fence.

Sweet peppers.

Planting Information: Sow your seeds directly into the soil (no need to start them indoors first), approximately 1 inch deep. Space your seeds about 2 to 3 inches apart, in rows about 12 to 18 inches apart.

Growing Time: 56 to 73 days.

Soil pH: The entire range of 6.0 to 8.0 is acceptable, but 6.0 to 6.5 (slightly acidic) is best.

Peppers

Types: The two main pepper classifications are hot and sweet. Both come in a range of shapes and a rainbow of colors: green, red, yellow, purple, and many more. Some popular hot peppers include jalapeños, poblanos, and habañeros (the hottest). Look for heirloom pepper varieties such as Fish (hot) and 'Purple Beauty' (sweet).

Requirements: Lots of sunshine is important for the success of pepper plants. A good quality soil, light and loamy, is suitable for peppers. They thrive on warmth and do not like frost. Keep them indoors until you're positive that the temperatures are sufficiently warm in the spring and there's no danger of a late spring frost and be ready to cover or pick your produce in the autumn if an early frost is imminent.

Special Considerations: Hot peppers are exactly as their name implies: hot. As you probably know, they are hot to eat, but they can also wreak havoc on your hands if you're not careful. Exercise caution when handling peppers and wear rubber gloves to protect yourself.

Planting Information: The fact that peppers need an ample growing season, coupled with the fact that they are very sensitive to frost, means that you'll need to extend your growing season in some fashion. Starting indoors is the best way to achieve this. Start the seeds in flats at a depth of ⅛ to ¼ inch deep. When the seedlings are about 2 inches tall, you can transplant them to larger containers, and by the time they are about 6 inches tall (approximately six to ten weeks after starting the seeds), you'll be ready to put them outdoors. Space them approximately 12 to 18 inches apart, allowing about 2 feet between rows. Peppers don't spread out the way tomatoes do, so it's easier to grow more plants in a smaller space.

Growing Time: 65 to 80 days.

Soil pH: 5.5 to 6.8 (moderately to slightly acidic).

Potatoes

Types: There's a potato to suit everyone's fancy—whether you'd like a quicker maturing or slower maturing variety or whether you'd like white, yellow, blue, red, or a host of other potato colors. Some varieties are noted for being excellent baking potatoes ('Yukon Gold', for instance); other varieties are exceptionally good for mashing, such as the 'Red Pontiac'.

Requirements: Like their fellow root vegetables carrots and parsnips, potatoes like a light, loamy soil that isn't too heavy. Avoid planting potatoes in the same location year after year; it's better to move them around to new locations.

Potatoes.

Special Considerations: Harvesting potatoes is fun, since you get to dig up these delightful starchy treasures from underground. If you wish to harvest "new potatoes," you can do so as soon as they have reached a sufficient size for eating. A new potato is not completely mature and not suitable for storage, but if you plan to eat them immediately, they are a treat. For the rest of your potatoes, you'll want to wait until they are fully mature before digging them up; watch the plant leaves, when they are turning yellow and brown in the autumn, it's probably time to harvest the potatoes. It's common to wait until after the first frost to harvest potatoes. Potato stalks and leaves are poisonous as are the white "sprouts" that grow out of the potato and any green spots on the potato. Do not consume any of those.

Planting Information: You don't have to wait around for the weather to warm up before you begin planting potatoes. To prepare your seed potatoes for planting, allow them to sit in a sunny area for a few days so that they begin sprouting. Then chop them up into smaller chunks, making sure that each chunk contains at least one eye, or sprout. Leave them for a couple of days, and then plant your potato "chunks" by placing them in a furrow about 4 to 6 inches deep, about 6 inches apart. Position the chunks with the eye facing toward the sky, then fill in the trench with a couple inches of soil. As the potato plant begins to grow and becomes visible above the ground, keep adding additional soil in a hill-like manner around the plant.

Growing Time: 80 to 120 days, depending on the variety.

Soil pH: 5.0 to 6.0 (acidic). Not over 6.0.

Radishes

Radishes are one of the earliest spring crops. They can be planted early, and their rapid germination and rapid growth mean that you could be harvesting radishes long before anything else.

Types: You can find red, white, or yellow radishes, but the most common type is the red cherry-shaped radish. Try the 'Early Scarlet Globe' variety if you're in a hurry (23 days) or 'Snow Belle' if you don't mind waiting a bit longer (30 days); the latter is a snowy white color.

Requirements: A loose, sandy soil will do nicely. Radishes don't require a lot of nitrogen, but they do appreciate full sun.

Special Considerations: If you're also planting carrots, consider planting radishes with them. The radishes germinate quickly and help farmers easily locate the slower germinating carrots.

Planting Information: Sow your radishes directly; there's no need to start them indoors first. Plant your radishes ¼ to ½ inch deep, spaced about 2 inches apart. The rows don't need to be very far apart; radishes are tiny, so you only need to allow a few inches between the rows.

Growing Time: Quick! As little as 20 days, although some varieties require up to 60 days.

Soil pH: 5.8 to 6.8 (moderately to slightly acidic).

Radishes.

Squash–Summer

Types: Many types of summer squash are available, including zucchini, yellow straight-neck, yellow crookneck, scallop (also known as patty-pan, a small round squash with scalloped edges), and many others.

Requirements: Sun and good drainage are vital for the success of summer squash. It requires a nutritious soil, and warmth is essential.

Special Considerations: Summer squash requires ample space, but it makes up for the space by providing ample production in return.

Planting Information: You can start your summer squash seeds indoors or out; I've had equal success with both methods, but their short growing season makes them a good candidate for planting directly into the soil. If you do start them indoors, be very careful when transplanting; their root systems are very sensitive. Indoors or out, plant your seeds 1 inch deep; if outdoors, plant in hills, with four to six seeds per hill (you'll thin them later).

Growing Time: Summer squash is quick to pop out the produce; figure on 42 to 50 days for most varieties. Don't assume that larger is necessarily better; many types of summer squash are best to eat when 6 to 8 inches long, although scallop squash are harvested when they are about 2 to 3 inches in diameter.

Soil pH: 6.0 to 7.5.

Summer squash (including zucchini).

Winter squash.

Squash–Winter

Types: The varieties of winter squash are abundant. Choose from, acorn, hubbard, butternut, spaghetti, buttercup, and pumpkin, among others.

Requirements: You need lots of space to grow winter squash. They like nothing better than to spread out and take over, so beware. In addition to ample space, you want a sunny location and a nutrient-rich soil.

Special Considerations: Don't be in any hurry to harvest your winter squash. Summer squash and winter squash are entirely different things—and winter squash needs plenty of time to fully mature.

Planting Information: You'll want to start winter squash seeds indoors, but be very careful when transplanting so that you don't disturb the roots. Plant the seeds 1 inch deep, in hills, and transplant with about 4 to 6 feet between seedlings.

Growing Time: Some hybrids have been developed for earliness, but you would be hard pressed to find any that grow in less than 70 days, and that's for a little acorn squash. Figure on at least 90 and up to 140 days for most winter squash plants.

Soil pH: 5.5 to 7.0 (moderately acidic to neutral).

Tomatoes

Types: Tomatoes are found in two main types: determinate and indeterminate. The differences are simple: determinate varieties are shorter, bushier plants that produce for a limited period of time (i.e., the growing time is precisely determined). Determinates are ideal for small farms where space is at a premium or for growing in containers. Indeterminate varieties grow and spread throughout the season and produce continually (i.e., the growing time is not precisely determined). They produce a higher volume of tomatoes but require much more space. Within both types, you'll find myriad colors and shapes: cherry tomatoes, 'Roma', egg-shaped, and the traditional round tomatoes. As far as colors, the rainbow is virtually limitless: green, white, yellow, purple, pink, red, orange, and even striped.

Here's another thing to remember: Tomatoes are found in heirloom and hybrid varieties. Many of the hybrids have been carefully developed for resistance to diseases and pests. You'll see letters after the names of hybrids, such as 'Better Boy (VFN)', which means it is resistant to verticillium wilt disease (V), fusarium wilt race 1 (F), and root-knot nematodes (N). If a variety is also resistance to fusarium wilt race 2, the designation is "FF," and a resistance to tobacco mosaic virus is indicated by "T." Heirloom varieties, while not exhibiting the resistance that the newer varieties boast, do have an advantage in the fact that you can save their seeds. Additionally, the flavor benefits are marvelous indeed!

Many people select the 'Brandywine' variety for its flavor, while others prefer 'Cherokee Purple'. Our favorite heirlooms are the 'Wapsipinicon Peach' and the 'Green Zebra'. Our favorite hybrid is 'Juliet'; it is prolific, tasty, and resistant to cracking.

Requirements: Sunshine and plenty of regular watering.

Special Considerations: For indeterminate varieties, you'll need tomato cages to hold up their sprawling and flourishing branches. Some people allow their indeterminate tomatoes to grow on the ground in a vine-like appearance, but you'll harvest higher yields if you use tomato cages, in addition to the fact that they keep your tomatoes off the ground, which results in cleaner tomatoes and keeps them out of the mouths of small rodents.

Planting Information: You can start your seeds indoors, or you can buy seedlings. Typically, it's tough to start seeds outdoors. Tomatoes want warm soil, so seedlings are the way to go. For seeds, start them in flats or peat pots, planting the seeds about ¼ inch deep (go ½ inch if you sow directly outdoors). When the seedlings reach 2 to 3 inches tall, transplant them to larger containers if you started them in flats; otherwise, wait until they reach about 6 inches before transplanting them outdoors (assuming, of course, that the weather is satisfactory).

It's wise to plant tomatoes using the trenching method, which doesn't necessarily have to involve a trench. Basically, it

Yellow tomatoes.

Hybrid tomatoes.

means that you plant the seedlings with a good portion of their stem underground. Some people plant them literally sideways, then curve the stem upward so that the top portion is above ground. This type of planting system allows your tomato seedlings to establish a strong foundation of roots, as new roots will grow underground all along the stem.

Growing Time: 70 to 85 days. Some hybrid varieties are known for their short growing times; for example, the 'Early Girl' matures in 52 days.

Soil pH: 6.0 to 6.8 (slightly acidic).

HAY

In its simplest definition, hay is grass that has been cut and dried. Nowadays, we add a third step—baling—to the sequence. The type of hay that you grow or buy will depend on your location and your animals' needs.

Grass hays come in several varieties including orchard grass, timothy, Bermuda, and brome. Alfalfa, clover, and trefoil are types of legume hay and contain more nutrition than grass hay. However, legume hays are very rich and may or may not be suitable for certain animals. You might opt for a grass/alfalfa mix in order to benefit from the increased nutrition and decreased richness.

Hay is the backbone of a winter feeding program and your livestock—your horses, cattle, sheep, rabbits, and goats—will depend on hay as their source of winter forage. Obviously, you'll want to provide your animals with the highest-quality hay, and you can recognize good hay by these characteristics:

- Good smell: High quality hay smells sweet and fresh, not earthy or musty.
- Low dust: Hay with a significant amount of dust may have been baled when it was too wet, or it may have been rained on while out in the field. Properly baled hay will contain minimal dust.
- Lots of leaves and few stems: This signifies that the hay was cut at the proper time; hay that is too mature will exhibit thicker, coarser stalks and has fewer leaves.
- Few weeds: Aim for hay that has minimal weeds.

If you can find a reliable hay source that produces quality hay, buying all of your hay can be a smart plan. But if you'd rather not be searching around for hay sources and trying to cobble together sufficient hay for winter, you might want to try taking control of your hay and producing your own crop.

Square bales are small enough to be managed by hand, which makes them a popular choice for many farmers and makes them easy for daily feeding.

Small square bales stack easily, as shown in this beautiful barn.

Producing your own hay has many benefits, but control is probably what many people appreciate the most. It's frustrating when your hay supplier suddenly decides to switch from hay to corn (it happens), or when you discover at the last moment that your hay supplier has sold your load of hay to someone else. If you make your own hay, you can control when you cut, how much you cut, and how it's stored. These three traits are very important to many livestock owners, so if you'd like to join the ranks of small-scale hay producers, let's take a quick peek at some of the necessary steps.

Putting Up Hay

In order to harvest hay, you'll need the following:

- A field of grass or a legume crop, such as alfalfa (if the field has been well-cared for and is free of weeds)
- A stretch of clear weather (at least three days without rain)
- A tractor and haying equipment, such as a mower or haybine, a hay rake, and a hay baler

To put up square bales, you'll want a hay wagon (or two), and a hay elevator is also helpful. You'll also want to round up as many helpers as you can find. When haying, remember two old sayings: *Many hands make light work* and *Make hay while the sun shines*. Both are important to keep in mind during haying season.

Putting up hay is a mixture of art and science, and a lot of both go into making the perfect bale of hay. Talk to farmers in your area for tips and advice. Ask lots of questions, and learn as much as you can before you start cutting. The specifics for haying do vary depending on several criteria, including your geographic location, the type of hay that you're cutting, and other circumstances in the Upper Midwest. It's fairly typical to cut hay on the first day, rake it into windrows on the second and third days (or sometimes just the third day), and bale on the third day. This allows the hay to dry thoroughly before baling. But many variables must be considered, and you'll want to seek advice on the best methods for your area.

Raking cut hay with a John Deere tractor prior to baling.

After cutting—on the second or third day—it's common to rake the hay into windrows.

Additionally, you'll want to invest in a hay moisture tester, a handy device that allows you to test the moisture content of your windrows and bales and helps you to avoid baling hay that is too wet.

Proper Storage of Hay

Storing hay is all about one objective: keeping it dry. Square bales should be stored indoors, stacked in a well-ventilated building. The ventilation aspect is extremely important, especially for newly cut hay that has not been fully cured. Uncured hay releases lots of moisture that must be allowed to escape in order to prevent mold or mildew in the bales. On fair weather days, try to keep doors and windows open to allow fresh air to enter and humidity to escape.

You'll often see round bales stored outdoors, but unprotected storage does result in an increased loss of usable hay. At the very least, round bales should be covered with plastic to prevent moisture from seeping into the bales; storing the round bales under cover is even better.

On the third day, the hay is baled. Here the people riding the hay wagon are stacking the hay as it comes off the baler.

A load of hay and straw is delivered to a barn. The gold-colored bales in front are the straw.

Large round bales of hay are stored under a shed roof to protect them from the weather.

SQUARE BALES? OR ROUND?

Hay is generally harvested and baled in two different ways: square bales and round bales. Square bales are actually rectangular shaped, and the hay is tightly compressed and tied into uniform shapes. These square bales weigh between 40 and 60 pounds, and they are ideal for stacking and storing indoors. Their small size is also ideal for feeding horses and small animals, as the average person can easily handle bales of this size.

Round bales are often used for pasture-feeding situations and are created by rolling the hay into a wheel-shaped stack about 5 feet wide and 5 feet tall. These bales weigh anywhere between 500 pounds to a ton, depending on the type of hay and the baler that's being used. Round bales cannot be moved by hand, so you'll need equipment (such as a tractor with a front end loader) to move the hay for feeding or storage. (Some people also choose to make a larger square bale that is comparable in weight to a round bale.)

ALL ABOUT STRAW

If you're harvesting grain, such as oats, wheat, or barley, you also have the opportunity to make straw. After the combine goes through and harvests the heads of the plants, the plant stalks are left behind. These stalks can then be harvested into straw bales. As with hay, straw bales can be square or round, depending upon your desired use. Straw can be very useful as animal bedding (it's my bedding of choice for mares with young foals), and many people also utilize straw in their gardens. Straw can also be a beneficial addition to a compost pile.

Even if you don't have a personal use for straw on your farm, you can always consider the possibility of offering your extra bales of straw for sale. You might be surprised at how much interest there is in obtaining straw. Construction companies in northern climates sometimes use straw to keep home foundations warm during construction; landscapers frequently utilize straw for outdoor projects; gardeners love straw for a variety of purposes; and it's easy to find many animal-related uses for straw. Straw is also popular as decorating items around Halloween and Thanksgiving. An inexpensive ad in your local newspaper or on Craigslist could yield more interest than you might expect!

COMPOST AND FERTILIZER

If you're going to invest the time and effort to grow small-scale crops, you'll want as many things in your favor as possible. Make sure that your soil is high quality and contains the appropriate nutrients and organic matter to support optimal growth. To achieve the best growing soil, you may wish to supplement it with compost and/or fertilizer. Let's explore the options and learn to make compost.

Compost

Generally speaking, compost consists of organic material such as dried leaves, seaweed, hay, grass clippings, eggshells, coffee grounds, and vegetable peelings, as well as manure.

A compost heap is low-key and simple; it's essentially an open pile of compost placed wherever is most convenient to you.

It's very important to select an appropriate place for your composting project. Look for an out-of-the-way place that is still convenient for daily use and then begin gathering your compost components.

If your farm's soil is less than ideal (for instance, pure sand), you'll want to improve it with a few tons of delivered compost.

Mix together the organic material you have—dried leaves, fresh grass clippings, rotten produce, kitchen leftovers, and the like.

Next, place a quantity of manure on your compost heap. Mix the pile thoroughly. Add more manure, then more organic material. Then let it sit.

In a few days, come back and stir things around. Add more organic material on a regular basis. If your compost appears too dry, try adding water, but if your compost begins to emit a bad odor, the mixture may be too wet. Hold off on adding any more water until the compost has a chance to dry out and the smell subsides. A bad smell can also indicate the presence of too much nitrogen in the compost. You can attempt to remedy the situation by adding dry leaves to your compost as this helps reduce the percentage of nitrogen.

So, how long will it take to turn your organic material into usable, valuable, unbelievable compost? The length of the process varies (it depends somewhat on the type of manure that you use), but figure on waiting at least 60 days before using your compost. The process of composting produces heat, which kills the bacteria in the manure, so in order to avoid potential harm, do not use fresh, uncomposted manure, and don't use partially composted manure. Only spread composted manure when the composting process is completely finished. You'll know your compost is ready to use when it turns dark and crumbly (sort of like the consistency of chocolate cake) and doesn't smell. The best time to add compost to your soil is late in the season—after harvest. Slowly work the compost into your soil, and then let it sit over the winter.

This compost has been aging for months, waiting to be applied next spring. Most any organic material may be added to the pile, but carbon and nitrogen must be in balance for the most efficient use of the materials.

THE BASICS OF MANURE

Got manure? If you have any type of livestock or poultry, the answer is a resounding Yes! You'd probably like to convert that pile of manure into composted fertilizer, so let's explore the varying types of manure:

Horse: Horse manure must be thoroughly composted before use. It's known as a "hot" manure and could burn your plants if it's not composted long enough before you use it.

Poultry: Like horse manure, this is another "hot" manure. It is slightly higher in nitrogen than horse or cow manure.

Cow: This type of manure is most likely what you will find at your local farm center—and although it isn't as nutrient-rich as some of the other types, it is certainly suitable for most situations.

Goat and Sheep: These types of manure are tidier than cow, horse, or poultry due to the fact that goat and sheep manures come in a convenient pelleted form. Don't let this convenience tempt you into spreading the manure prior to composting, however. This is not recommended, and it's always better to compost first.

Rabbit: Rabbit manure contains high amounts of nitrogen and phosphorus (more than any of the other manure types mentioned). Some people consider rabbit manure to be nature's ideal fertilizer. Remember: Don't spread before composting; always let it rot first.

A few tips to remember: Do not use dog or cat manure in your compost. These types of manure can contain harmful parasites and diseases. And when adding kitchen scraps, try to steer clear of fatty, greasy, oily foods and meat scraps, because they may attract rodents, which is not a good idea.

Fertilizer

Fertilizers are an excellent source of three essential nutrients: nitrogen, phosphorus, and potassium. These nutrients together are commonly referred to as NPK (N stands for nitrogen, P stands for phosphorus, and K stands for potassium). The numbers on fertilizer bags represent the percentage that it contains of each nutrient, such as 10-10-10, which translates to 10 percent nitrogen, 10 percent phosphorus, and 10 percent potassium.

You want to choose a fertilizer that contains the closest amounts to what you need in your particular soil. For instance, if soil testing has proven that your soil is too acidic, you want a fertilizer with a higher concentration of nitrogen—perhaps a 15-10-10. Or you can select

Autumn leaves are a key ingredient in most mulching and composting systems.

individual products for your soil if you have a specific need. For instance, soil that needs an increase in phosphorus could benefit from the addition of some bone meal. If potassium is lacking, try kelp meal. Soil that is too acidic could be treated with lime, and sulfur is helpful to lower the pH of your soil.

You can also look for slow-release fertilizers that are designed to provide small, continuous amounts of fertilizer of a lengthy period of time.

Organic and inorganic (synthetic) fertilizers both deliver the same content (nitrogen, phosphorus, and potassium), but their origin and impact on the soil differs greatly. Organic fertilizer comes from natural sources (plants or animals), and synthetic fertilizer is manufactured. Synthetic fertilizers work very quickly, but they don't add any organic matter, which is the stuff that makes your soil rich and fluffy. Many people prefer to use organic fertilizer such as composted manure or compost made from kitchen scraps and plant debris. If you have horses, rabbits, chickens, or other animals on your farm, you'll have an ample supply of manure that you can turn into a rich compost for your crops. If you go this route, you're on your way to farming organically, which has added environmental and economic benefits.

Chapter 3
Livestock

In this chapter, we'll introduce you to all types of livestock, from rabbits to goats to cattle. You'll also learn about chickens and other types of poultry, and you'll meet the smallest farm worker: the industrious honeybee.

Raising livestock is a fairly simple activity. Food, water, and shelter are the basic ingredients needed to sustain the life of any farm animal, whether beef cattle or baby bunnies. Regardless of the breed or whether the animals are large or small, you'll need to consider four factors

Raising livestock isn't difficult: you need food, water, and shelter for any farm animal—large, small, or in between.

when deciding what you can raise on your farm: the amount of space or acreage, feed supply, the time you can commit, and the available money. To be successful, you will need to put some thought into your goals and figure out what costs are involved. Planning ahead helps you avoid unpleasant surprises.

Large animals, such as beef or dairy cattle, typically cost more to purchase than smaller animals like pigs or sheep. This formula is based on the market value of a live animal, which is determined by the volume of meat that can be derived from the carcass for processing.

Other factors are involved in the value or price of an individual animal, including popularity of the breed, pedigree or ancestry performance, and the supply and demand of a given breed.

The amount of acreage you have available will be one of the factors that may influence the kind and number of animals you raise. In general, the smaller the animal, the less space you'll need. Rabbits and chickens need less space than a pig. Sheep, goats, and pigs take less space than cattle. Beef cattle require fewer facilities and equipment than milk cows.

All animals have some requirements in common. An in-depth look into each species can be found in the *How to Raise* series of books (nine in all) and other livestock books by Voyageur Press (see Resources).

BREEDS FOR YOUR FARM SITUATION

Each farm is unique. Animal breeds best suited for your purposes may be totally different from that of your neighbor. The region of the country where you live may influence your choice. Some breeds are able to tolerate different climates better than others. Besides climate considerations, breeds also differ in traits, such as mature size and calf growth rate. These factors impact nutritional requirements and eventual production costs. Breeds also differ in growth rate, reproductive efficiency, and maternal ability. Choose an animal breed that matches your goals.

Most all livestock breeds offer registry programs. Registries are necessary if you choose to raise pedigreed animals or purebreds versus unpedigreed animals or grades. The parentage of purebred animals can be substantiated, and the tracking of their lineage may allow them to participate in purebred programs, such as shows and fairs. Being listed in a registry may also open up a market for seedstock purposes.

The purebred versus commercial markets do not need to be competitive as they only differ in approach and end goals. Raising quality livestock can be accomplished whether your animals have a documented ancestry or not.

Breed Variations

The different breed associations can provide information to help you. Although individual animals may differ, some general guidelines and variations apply to different breeds.

COMMON AND HERITAGE ANIMAL BREEDS

BEEF

Most Common	Heritage
Aberdeen-Angus	Florida Cracker/ Pineywoods
Brahman	Kerry
Charolais	Randall/Lineback
Chianina	South Devon
Devon	Dexter
Galloway	British White/White Park
Hereford	
Limousin	
Polled Hereford	
Red Angus	
Red Poll	
Santa Gertrudis	
Scottish Highland	
Shorthorn	
Simmental	

DAIRY

Ayrshire	Dexter
Brown Swiss	Dutch Belted
Guernsey	Lineback (American)
Holstein	Milking Devon
Jersey	
Milking Shorthorn	
Red and White	

GOATS

Dairy	Fiber	Meat	Heritage
Alpine	Angora	Boer	Arapawa Island
LaMancha	Cashmere	Kika	Golden Guernsey
Nubian	Colored Angora	Myotonic	San Clemente
Saanen	Nigora	Savanna	
Sable	Pygora	Spanish	
Toggenburg			
Miniature: Pygmy			

SHEEP

Wool/Meat	Dairy	Heritage
American Blackbelly	East Friesian	Cotswold
Barabados Blackbelly	Lacaune	Jacob
Border Leicester		Karakul
Cheviot		Leicester Longwool
Columbia		Navajo-Churro
Corriedale		St. Croix
Debouillet		Scottish Blackface
Delaine Merino		
Dorper		
Dorset		
Hampshire		
Katahdin		
Lincoln		
Perendale		
Polypay		
Romanov		
Romney		
Shetland		
Shropshire		
Southdown		
Suffolk		
Targhee		
Texel		
Tunis		
Wensleydale		

SWINE

Most Common	Heritage
American Landrace	Guinea Hog
Berkshire	Hereford
Chester White	Lacombe
Duroc	Large Black
Hampshire	Mulefoot
Spotted Poland China	Red Wattle
Yorkshire	Tamworth

BEEF AND DAIRY CATTLE

You can raise beef or dairy cattle to market animals for meat or producing milk and dairy products. Both have their own rewards. Large animals can have several additional challenges when compared to raising small animals such as sheep, goats, pigs, or poultry, but they are not so great that you cannot succeed. Quite the contrary, a growing niche market has been expanding by those raising cattle on small farms and developing a marketing plan to sell quality meat or milk and dairy products. Today's consumers are seeking high-quality food from animals grown under wholesome conditions that can in some way connect these consumers to the land. This connection can provide a strong link with new customers. Cattle are quite self-sufficient and require the same good management practices used with other farm animals.

Housing

Beef cattle require little in the way of housing. Often they prefer being out in the elements provided they have access to water and at least an opportunity to shelter in very inclement weather. Dairy cattle require more housing because they will be milking and need shelter in extremely cold, wet weather. They are not as durable as beef cattle because their purpose is different, although they can be raised to produce meat too. The most important difference is that a milk cow has teat structures that can become frozen during very cold weather because of the limited blood flow through the tissues. Dry, clean bedding is essential for a dairy cow; beef cattle can survive on cold, hard ground.

The South Devon breed is an English native that exhibits a good growth rate, excellent carcass quality, and are naturally polled (no horns). Their red color matches that of the Red Angus but the two breeds are genetically far apart.

TOP: A beef herd does not need to be large in number to provide meat for you and your family. Cattle and other livestock can utilize areas difficult to mechanically harvest.

ABOVE: Jerseys are a small-size dairy breed. They produce milk high in butterfat and protein, excellent for cheese and butter making.

One of the most important aspects of housing is to provide a place where your beef or dairy cows can escape the cold, bitter winds and rains. A shed, lean-to or other building such as these will often provide a suitable stable.

Feed Requirements

Cattle are ruminants and can live solely on pastures and grasses if that is what you have available. Dairy cows often require more energy for the production of milk but this is not a necessary consideration if you are only producing milk for you and your family. If you are producing for a large customer base then you will want to sustain an optimum level of production, which may include supplemental feeds and proteins.

As a general rule, large animals such as beef cows and mature dairy cows require more pounds of feed per day than smaller animals such as calves and yearlings because of body size. A larger body requires more energy and protein for production of muscle or milk than a smaller body. Likewise, differences between dairy breeds also dictate that a Holstein will typically consume more feed than a Jersey. The total amount of feed you need to plan for is determined in part by the cow breed you choose.

Beef cattle feed requirements are often calculated as a cow with a calf by her side. As a rule of thumb, one cow-calf pair will consume about 4 percent of the cow's body weight in dry matter (grass and/or dry hay) each day. A dairy cow typically will consume between 2 to 2.5 percent of her body weight in dry matter each day. These percentages hold true whether she is a large cow or a small cow. Only the total volume will change.

These figures can be used to calculate the daily, monthly, quarterly, or yearly amounts of feed you will need to provide to each animal in your herd, whether it comes from grass and pastures or purchased feeds.

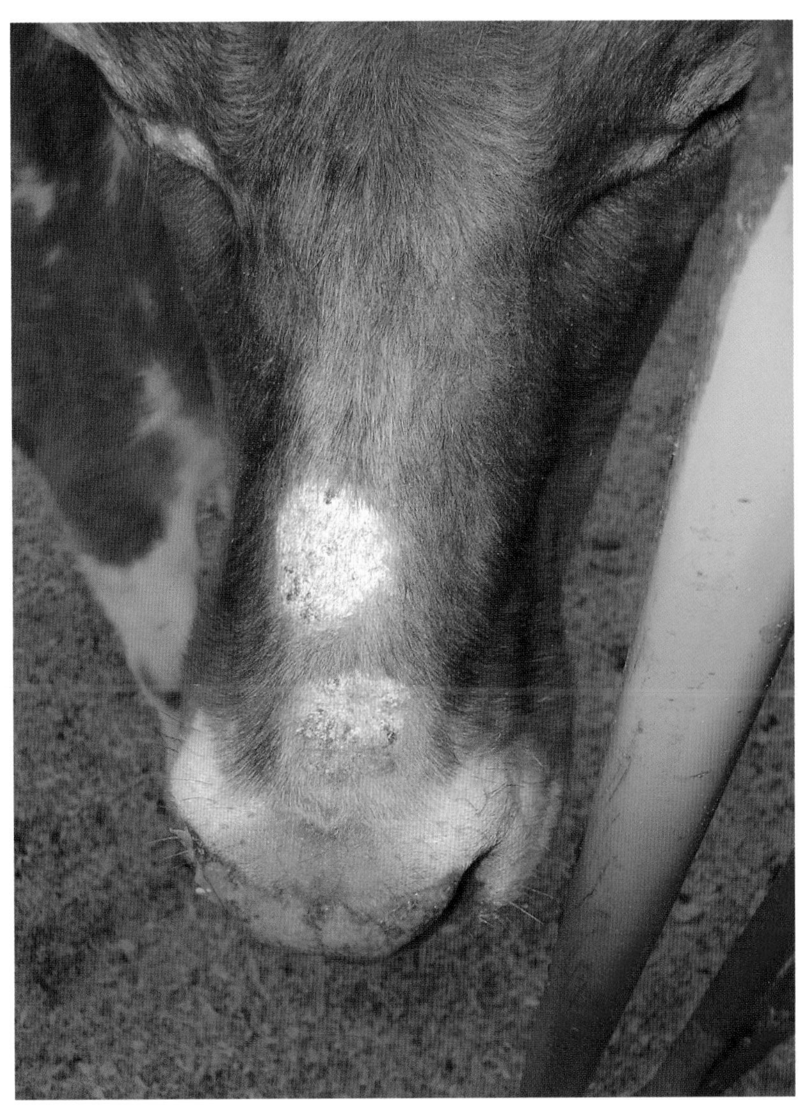

Ringworm is a highly contagious skin infection that can affect any species. Though it looks tender, there is little pain involved, although animals affected typically rub it to relieve the itching. Be sure to use good sanitation practices when treating ringworm.

Start-Up Costs

The cost of purchasing a beef cow or a dairy cow can vary depending on several factors, such as if a calf is included, the quality of the animal, the age, and its health status. Cow-calf beef pairs cost a little more than a cow purchased singly. Dairy cows are often priced on their milking ability. The higher production she has, the more likely the price will increase. Younger animals typically cost more than older ones because they have a longer expected life than an older animal. Regardless of price, it is never a good option to buy an animal that does not seem physically sound or has obvious health problems.

Market conditions may change from season to season and year to year, but a single beef cow may cost $850. You may need to add another $150 if there is a calf with her. A single beef calf may cost $140 or more. The price of a dairy cow may change during stage of lactation. If she is soon to have a calf, the price is often more because of the added bonus of a potential offspring. The price may drop toward the end of her lactation because most of her year's milk production has already been harvested. If you buy a late-stage lactation dairy cow, make sure that she is pregnant so that you don't have to wait another nine months for a calf. A typical dairy cow may cost $900 for smaller breeds such as Jerseys, or $1,200 for a Holstein or Brown Swiss. These numbers can then be multiplied by the number of animals you want for your beef or dairy herd.

Acreage Needed

The amount of acreage needed for a beef or dairy herd is intimately tied to the number of animals you want to raise. Typically, a beef cow-calf should be allotted about 1½ acres for the year. A dairy cow may need up to 3½ to 4 acres per year. This figure includes all the forages produced on that acreage during the year and not necessarily as a space for movement.

Fencing Requirements

Large animals such as beef and dairy cows need good fencing. Woven-wire fencing and three or four strands of barbed wire are essential for permanent or perimeter fences to keep your cows from escaping into the neighbor's fields or onto roadways. Temporary or movable wires and posts can be sufficient for paddock layout inside your perimeter fencing. Cattle fences should be strong enough to keep them from breaking down under pressure of large bodies pushing against them and solid enough to last for several years. Permanent fences should be at least 4 to 4½ feet at the top to reduce the chance of any cow trying to reach or push over. If you are using wire strands, they should be set close enough, say 6 to 8 inches apart, to deter a cow from sticking her head through and pushing against the wires to reach something of interest on the other side. Good fence management is good husbandry.

Marketing

Beef and dairy cows can provide meat for a market. Dairy cows have the added bonus of providing milk for making a variety of dairy products such as butter or cheese. Home-grown beef is gaining popularity with consumers and can provide an additional source of income if you develop a label under which to sell your meat. Rules and regulations govern the selling of meat

For a livestock operation your equipment needs may include a skid-steer loader. These are stable implements that can be used for many jobs on your farm.

American bisons are known for their lean meat. They need strong fencing and calm handling. Their herding instincts are very strong, so they should not be separated from the herd for long periods.

and/or milk from your farm, but these ordinances can be followed with appropriate advice and help from local, county, or state agencies that oversee such activities.

BISON

Bison are not much different or difficult to raise than other cattle. Although often associated with the wild ranges of the western United States, bison increasingly have become privately owned through bison ranching in many areas.

Feed Requirements

They can be stocked at about the same rate as cattle or even a little higher because they are efficient grazers. They will spread out more evenly over the pasture than cattle and can survive on marginal pastures that would starve other cattle.

Bison will travel about 2 miles per day to graze and can be kept content through rotational grazing. Because of their strong herding instinct, it is best to maintain a herd size of about 10 to 15 animals. A lone bison will tear down fences or jump over them to get to other cattle, horses, or even sheep in order to satisfy its need for a herd.

Start-Up Costs

The start-up costs for purchasing bison are comparable to other cattle, although males may be a little more expensive. The considerations differ if you are purchasing a live buffalo for meat or for breeding. Breeding stock can vary between male and female, and usually varies among calves, yearlings, two-year-olds, and older animals. Heifer calves can sell for $600 to $2,500.

Meat bulls can cost from $1,200 to $3,000. Breeding bulls have sold from $2,500 to $10,000. Bison more than two or two-and-one-half years of age typically don't sell as well because of the desire for younger animals.

Like other cattle, bison need a large pasture for grazing, companionship with other bison, and an unlimited water supply. They have a strong herding sense, and it is difficult to keep one or two separated from the rest of the herd as a lone bison will do almost anything to get back with the group.

Bison generally need less hands-on management than domestic cattle because of their long history in the wild. This independence also makes them only semidomesticated with wild tendencies lying just below the surface. They can become temperamental, and you will need to use low-stress handling techniques. Be aware that both males and females have horns, and they can be skillfully used in any situation.

Fencing Requirements

Bison need more durable fencing because of their wild instincts as well as their massive force. Strong, tall exterior fencing is an absolute requirement before bringing bison onto your farm. They can sustain speeds between 35 and 40 miles per hour for several miles, and at full impact can lay waste to a fence very easily. A typical adult can stand about 6 feet tall at the shoulder and weigh more than 2,000 pounds. In addition, they have been known to jump 5 to 6 feet high at a standstill. Work bison slower than you would domestic cattle, and separate individuals from the herd as little and as calmly as possible. Always be alert for surprises with bison.

Bison have the ability to prosper in any type of environment and climate. They can thrive in extensive heat and humidity and in wind chills of minus 100°F. With this range of adaptabilities, bison can settle into most any surroundings and be very good at fending for themselves in any type of weather. Limited shelter is needed when compared to other cattle.

SHEEP

Raising sheep has several advantages compared to raising other livestock, and most of these advantages relate to the investment of money and time. Sheep are very self-sufficient and after you attend to the lambing season, they often don't require your presence. Good management, however, will require you to maintain a consistent surveillance to ensure they remain safe from predators and free of injury.

Housing

Sheep do not require extensive housing or equipment to be raised in a humane and productive manner. Often a shed, lean-to, hoop house, Quonset-style hut, or a well-ventilated building is all that is needed. Conversions of existing buildings can adequately house your sheep.

Purchasing healthy animals at the start is the best way to maintain a healthy flock or herd over time. Healthy animals are easier to raise because they grow better by utilizing feed more efficiently, and they cost less to maintain.

Feed Requirements

Sheep are ruminants and make efficient use of grass pastures, which decreases the amount of supplemental feeds required daily. The nutritional requirements for ewes increase during late gestation and lactation compared to the maintenance and early gestation periods. For a typical 150-pound ewe, you should provide daily nutrient requirements in 3.5 to 4.0 pounds of feed. Of this amount, the overall fiber content consumed should be a minimum of 40 percent or about 1.5 pounds but can increase to 4 pounds when supplemented with grass pasture or dry hay. Feeder lambs can be fed 2 to 3 pounds of grain and 4 to 6 pounds of hay per day, plus mineral supplements. Free choice water should always be available, even in cold months.

Start-Up Costs

The cost of purchasing sheep may be dependent upon the breed desired, its availability, the age of the animal, and in some cases, the rarity of the breed. Generally, a typical ewe more than one year of age will cost between $100 to $200, while a ram can cost twice that or more. A healthy ewe should live for several years, so this cost, averaged out over, say, four years, makes their cost quite reasonable.

Viral illnesses and vaccinations are typically handled through injections using a sterile syringe. There are many shapes and sizes of syringes and needles and you should become familiar with their use. Use needles only once and then properly dispose of them. Be sure to read and understand all label directions and withholding times for the products.

Windbreaks are usually located near fences and offer protection from the elements. You can also plan and plant windbreaks after fences are built.

Acreage Needed

One advantage of raising sheep is that you can do it with limited acreage. A 150-pound ewe needs between 1 and 2 acres of pasture per year. Sheep can be raised in confinement areas, but then all their feed will need to be brought to them.

Fencing Requirements

Woven-wire fences are better for sheep than barbed wire because they reduce the chance of injury and damage to the wool. Entanglement in wire by ewes and lambs is also reduced. Although some sheep can clear low hurdles, a fence that is between 26 and 32 inches in height should be sufficient. The most important fence you have is the perimeter fence as that is the last line of defense if they should escape from their enclosures or pastures. A sturdy fence will prevent them from entering highways and neighboring fields.

Marketing

Sheep provide meat, wool, and in some cases, milk for yourself or to be sold. Meat from sheep up to one year of age is referred to as lamb and is usually taken from a carcass weight of between 15 and 65 pounds. The wool from one sheep is called a fleece, and it may yield up to 8 pounds of wool. It can be sold to brokers or transformed into a product that can be used by hand spinners, knitters, and weavers. Sheep milk can be used for making cheese. Small-scale cheesemaking can be learned, and with proper equipment, it may become a part of your farm business. Sheep's milk cheese is gaining in popularity with consumers as a nutritious alternative to cow's milk cheese.

Dairy sheep breeds are used primarily to produce milk that can be made into cheese. The East Friesian produces the highest volume of milk per year of any sheep milk breed.

GOATS

Goats can be raised for meat, milk, hair, or the simple pleasure of having them as companions. Raising goats is similar to raising sheep. Their housing needs are typically the same. They can be raised in sheds, barns, a lean-to, or hut as long as they have access to dry bedding and good ventilation, particularly in cold weather.

Feed Requirements

Goats used for meat production will have slightly higher daily feed requirements because you are trying to increase their muscle mass to increase their weight. An increase of this nature requires more protein supplements and high-quality hay, although the animals can still be placed on pasture. Typically, a doe requires 2.5 to 4.0 percent of its body weight in feed each day. For example, a 100-pound doe will need about 3 to 4 pounds of feed each day in some form. Because a goat is a ruminant, it needs grass or hay rather than a full-feed of grains.

In general, you'll need one to two acres of pasture per goat, in addition to supplements and feed.

A goat's shelter needs are basic: protection from the weather, dry bedding, and good ventilation.

Start-up Costs

Similar to sheep, the cost of a doe or buck is often dependent upon whether it is used for meat, milk, or fiber. Typically, a doe will cost between $100 to $200, depending on such factors as availability, breed or rarity, and use. A buck may be two to three times that cost, depending on its age. Goats registered with a breed association may cost a little more.

Acreage Needed

Goats are similar to sheep in their pasture requirements. Typically, each goat will need access to 1 to 2 acres when considering the feed requirements on a yearly basis. They can do with less than this but will then need access to stored or purchased feeds.

Install a goat-proof fence before you bring the kids home.

Fencing Requirements

A goat's agility presents fencing challenges. Woven wire panels and chain link are good options for fences. Hog panels are 3 feet high and have a narrower spacing at the bottom than the top. Panel fencing is good for kid pens since the kids are too small to climb over or through them. Cattle wire panels or woven wire fences of 4 feet in height are hard for adult or adolescent goats to jump.

Marketing

Meat goats are in demand by many ethnic groups. About 70 percent of all red meat consumed worldwide comes from goats. Goats may be marketed at stockyard auctions, sold as live animals privately, or butchered and processed by you. Goat milk can be sold as fluid milk or processed into cheese for yourself or others. You can become licensed to produce goat cheese and develop a market for your own label.

PIGS

Raising pigs is a fairly simple farm activity, and like other species, pigs are fairly self-sufficient from an early age. They have endeared themselves to small-scale farmers because they grow fast on simple diets. Each litter can produce between 8 to 10 piglets, so they can replenish themselves quite readily. Small-scale pork production can be an option because it requires minimal inputs.

Housing

Pigs can be raised outdoors in fields or pastures where they can roam in all kinds of weather and still thrive. Or, they can be raised completely indoors with good ventilation. Like other species, pigs need to get out of bitter winds and cold, wet weather conditions when required. Sheds, huts, a lean-to, and barns can provide adequate shelter, but they need to be accessible to clean out. Pigs require shade in hot, sunny weather because of their inability to sweat. They need areas where they can stay cool. Pastures with trees are good areas to let pigs roam in summer.

Start-Up Costs

Initial costs generally depend more on age and weight rather than the breed, but this formula also may be affected by availability and breed rarity. Smaller animals cost less than larger ones but generally average out about the same on a per-pound basis. Feeder pigs between 40 and 50 pounds may sell for $35 each. Breeding sows can usually be purchased for about $150 to $200 each, boars for $200 to $250 each, and gilts for $150 to $200 each. These prices reflect that commercial pigs and animals registered with a breed association may cost more because of their value as breeding stock.

Pigs reproduce rapidly: a typical litter is 8 to 10 piglets.

COST AND ACREAGE REQUIREMENTS

	Cost of Purchase/animal*	Acreage Needed**	Forage Requirements***
BEEF/DAIRY	$800–$1,200	5–8 acres/cow-calf	4 percent body wt.
BISON	$1,500–2,000	5–10 acres/cow-calf	4 percent body wt.
GOAT/SHEEP	$100–200 (does/ewe) $200–350 (buck/ram)	1–2 acres/150# ewe/yr.	2.5 percent body wt.
PIG	$35–40 (feeder) $150–$200 (sows/gilts) $200–$250 (boars)	4 lbs. dry grains/day 8 lbs. dry grains/day 10-12 lbs. dry grains/day	Choice Choice Choice

* Approximate cost. Heritage breeds may cost more.

** Only forage calculated. Grains and concentrates not included except for pigs.

*** All animals except pigs can subsist fully on forages. Pigs need grains, being monogastric. All animals need unrestricted access to water daily.

Be sure to provide pigs with shade during the summer, as they don't have the ability to cool off by sweating.

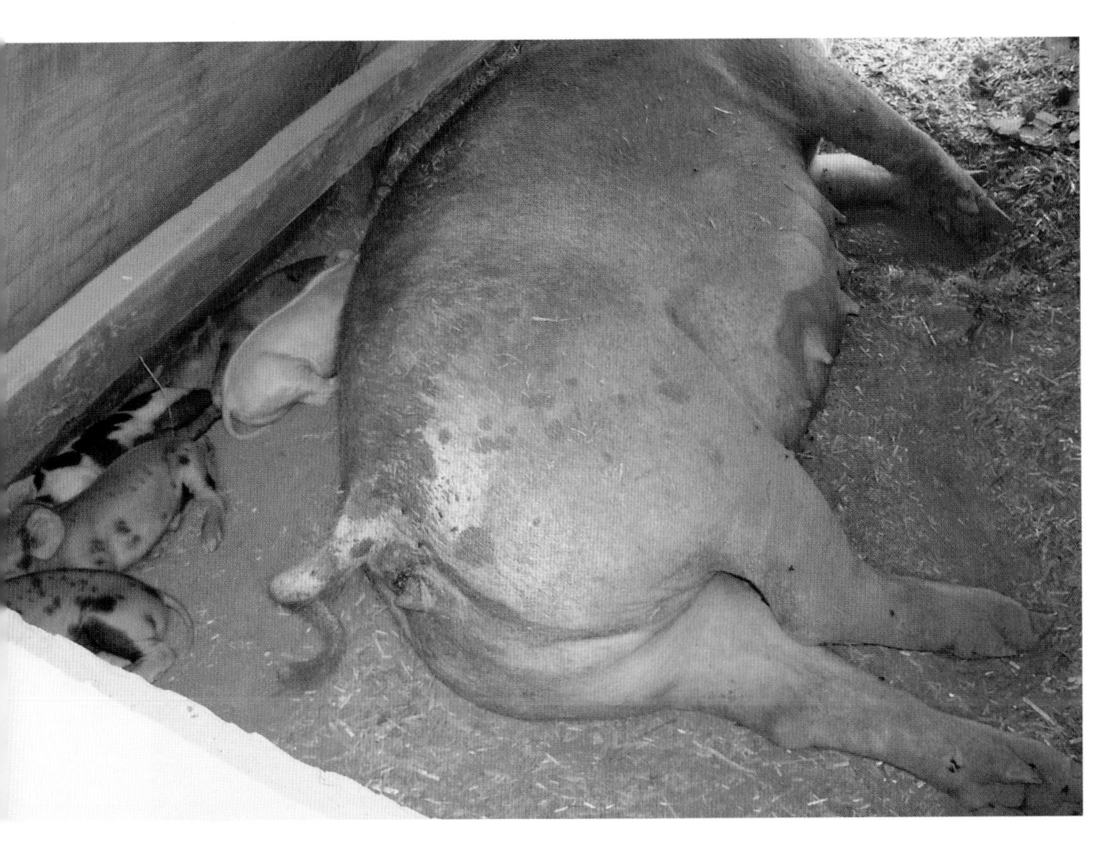

It's best to plan for farrowing (the time when a sow gives birth) to take place in the spring or fall, as piglets are vulnerable to extreme temperatures and require much more care to survive very cold and very hot weather.

Feed Requirements

Feed costs will be the largest expense during the growing period between purchase and market, or between birth and market. You can choose to raise your own crops or purchase their feed. Typically hogs require between 3 and 5 pounds of feed per pound gain through the three phases of production: nursery, growth, and finishing. Although pigs are not ruminants, they can utilize grass and pasture as a small portion of their daily diet, about 3 to 10 percent. Most of the feed eaten by young pigs is from cereal grains and protein supplements that are used for growth and maintenance. Mature pigs primarily use feed for maintenance and reproduction. Feeder pigs will eat up to 4 pounds of dry grain per day; sows and gilts will consume twice that. Boars can eat as much as 10 to 12 pounds of dry grain per day and more during the breeding season.

Acreage Needed

The amount of space needed will depend upon the size of the pig and, if a sow, upon whether or not she has a litter of piglets. Outdoor housing requirements for bred sows or a boar are between 15 and 20 square feet. For a sow with a litter, the area needed is about 30 to 40 square feet.

Fencing Requirements

Pigs will test your fences as they root and roam around the pasture or farm. Good fences are important. Woven wire fencing is one of the best materials to enclose pigs because it has smaller spacing at the bottom than the top

Pasture grass can provide a small portion of a pig's diet (less than 10 percent).

and discourages them from trying to push through. Perimeter fences are the most important to keep pigs from escaping your farm, and you should first work to make them sturdy. Wire panels are good for small enclosures. Because pigs cannot jump very high, a 3-foot wire panel fence will handle most of them. Electric fencing is an option, but it must be monitored to make sure it is working properly.

Marketing

Like other farm animals, pigs can be marketed in several ways. When they reach a sufficient weight, generally between 180 and 200 pounds, they can be sold at a stockyard or they can be butchered and processed into a value-added product with your own label. The latter carries certain requirements and an adherence to state and/or federal regulations but is a viable alternative for all your work in raising them. Or, they can be butchered for your family's use.

HORSES

For many people, a farm just isn't complete without a horse—or two, or ten. In addition to their utilitarian uses around the farm, horses provide companionship in a way that is unmatched by other types of livestock, which adds to their appeal.

A calm, easygoing horse is a pleasure to have around the farm. Look for a horse with a good disposition and no vices.

Getting Started with Horses

When you're just starting out, you'll want to evaluate your future plans for your horse. Will you be using the horse as a trail mount? A competitive show horse? A broodmare? A driving pony? A work horse? Consider your plans and requirements and create a wish list that incorporates these characteristics.

Some of the best advice you can receive is this: don't buy the first horse you see. It's all too easy to fall in love with the first pretty pony or stunning stallion that trots by. Resist the urge to make a snap decision. Instead, take your time and look at a number of horses before making your decision.

Health and temperament are absolutely paramount. You'll want a horse with a kind and gentle disposition that is not suffering from illness or injury. The difference between buying an untrained two-year-old colt and a fully trained twelve-year-old gelding is great. If you don't have the experience to take that untrained two-year-old colt and pursue his training, then you'll definitely want to search for a horse that is already trained.

HORSE LINGO

If you're just getting your feet wet with horses, you'll want to acquaint yourself with some of the terminology that you'll encounter in the horse world.

Mare—a female horse over the age of 3

Stallion—a male horse over the age of 3

Gelding—a castrated male horse of any age

Foal—a baby horse

Weanling—a foal that has been weaned (usually refers to foals between the age of 6 months and 1 year)

Yearling—a horse between the ages of 1 and 2

Filly—a female horse under the age of 3

Colt—a male horse under the age of 3

Pony—an equine that is less than 14.2 hands (it's a common misconception that a pony is a baby horse—it's not)

BREEDS

There are dozens of breeds out there from which to choose. Let's take a quick peek at some of the more common horse and pony breeds:

American Quarter Horse—America's favorite horse; the versatile Quarter Horse is the most popular breed in America.

Thoroughbred—Best known for their success in racing, the Thoroughbred originated in England and is an athletic yet elegant equine.

Arabian—Stunningly beautiful and rich with heritage and history, the Arabian is hugely popular in America.

Appaloosa—Noted for its striking range of color and markings, the Appaloosa is a distinctive stock-type breed.

American Paint Horse—Similar to the Quarter Horse in type and popularity, the American Paint Horse boasts the addition of overo, tobiano, tovero, and sabino coat color patterns.

Morgan—Developed in North America, the lovely Morgan is pleasing to the eye and a versatile performer.

American Saddlebred—A gaited breed, the American Saddlebred is a popular show horse under saddle as well as in harness.

An Appaloosa grazing.

Welsh Mountain Ponies are often used for pleasure driving.

American Standardbred—Still recognized as the fasted trotting horse in the world, the American Standardbred is a talented breed with natural endurance.

Tennessee Walking Horse—Best known for its famous four-beat running walk, Tennessee Walking Horses also have excellent temperaments.

American Miniature Horse—Beauty and brains in one diminutive package! The American Miniature Horse is the smallest of equines but maintains horse-like proportions.

Friesian—Majestic and powerful, the black Friesian is stunning to behold.

Missouri Fox Trotter—The gaited Missouri Fox Trotter performs the flat foot walk, the fox trot, and the rocking horse canter.

Welsh Ponies and Cobs—With four types to choose from—the Welsh Mountain Pony, the Welsh Pony, the Welsh Pony of Cob Type, and the Welsh Cob—there's truly a Welsh for every member of the family.

Shetland Pony—Noted for its hardiness, the Shetland Pony is found in two types: the classic Shetland, which stands 42 inches or less, and the modern Shetland, which contains Hackney, Arabian, and Thoroughbred blood and can be as tall as 11.2 hands.

Pony of the Americas—Appaloosa coloring in a pony package! The Pony of the Americas (known as the POA) is a talented, versatile pony with plenty of athleticism.

Belgian—One of the most popular draft breeds in America, the Belgian is noted for its strength and gentle temperament.

Percheron—A draft breed hailing from France, the Percheron is usually found in gray or black.

Clydesdale—Named for the River Clyde in Scotland, the Clydesdale is a draft breed that's most commonly associated with the Budweiser Clydesdale hitch.

Shire—Developed in England, the Shire horse is named for the Shires (counties) in England and is one of the tallest draft breeds.

Shire, a type of draft horse.

DONKEYS AND MULES

Donkeys are sometimes used as livestock guardians for smaller farm animals, and mules are sometimes used as draft animals. Donkeys and mules have long been noted for their ability as pack animals, and although they resemble their horse cousins in many respects, they also boast distinctive characteristics. At the top of the list: their exceptionally long ears and relatively sparse manes and tails. They come in a wide range of sizes—from diminutive miniature Mediterranean donkeys that stand less than 36 inches tall all the way up to draft-size mules. Don't overlook donkeys and mules as a possible addition to your farm menagerie, but don't be thinking you are going to save a lot of money by choosing one instead of a horse: on the contrary, quality donkeys and mules can carry a hefty price tag.

Donkeys and mules are easily identifiable by their distinctive ears. They are intelligent creatures that are enjoyed by many farm families.

SIGNS OF A HEALTHY HORSE

- Alert, good appetite, bright and energetic
- Normal rectal temperature of 99.5 to 100.5 degrees Fahrenheit
- Resting heart rate of 25 to 45 beats per minute
- Properly hydrated
- Normal capillary refill time
- Normal gut sounds (gurgling from deep inside his stomach)
- Normal drinking and eating habits
- Manure that is of normal consistency (not hard and dry, or sloppy and wet)
- Sound (not lame)

Correct Conformation

The importance of conformation is sometimes overlooked by novice horse owners, because it sometimes seems less important than other equine characteristics like good dispositions and health and soundness. But correct conformation—that is the balance and structure of the horse—is one of the main elements of a horse's ability to perform well in athletic pursuits. If you ever wish to show your horse, you'll find that correct conformation is paramount. If you ever wish to breed your horse, you'll want his or her conformation to be as correct as possible—individuals with major conformation faults should not be used in a breeding program.

A horse with correct conformation is balanced. This means that the front of the horse should be roughly equivalent to the mid-section and the hindquarters. The neck should be lengthy and should flow smoothly into the withers and into the back. Shoulder slope is another important characteristic, as it determines the length of a horse's stride. Shoulders that slope produce freer movement, while straight shoulders restrict the horse's ability to achieve a long, smooth stride.

Short backs are more desirable than long backs, since short backs are fundamentally stronger. Hindquarters should be deep with the croup not being too steep. Legs should be straight with hooves that do not turn in or out and pasterns that are of moderate slope.

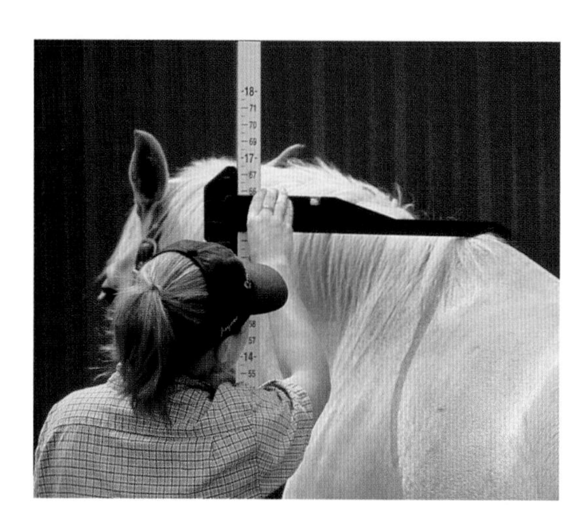

WHAT ARE "HANDS"?

The height of a horse is measured in hands. One "hand" is equal to 4 inches, and the term stems from the olden days when horse owners would gauge the size of a horse based on the width of a man's hand. Nowadays, horse measurement is a bit more precise—we have specialized measuring sticks that accurately measure a horse's height in inches, but we still convert those inches to hands.

WHAT YOU WANT AND DON'T WANT IN A HORSE

Characteristics to Seek:

- Can be safely turned out to pasture with other horses
- Can be bathed, clipped, tied, loaded, and dewormed without difficulty
- Doesn't chew wood
- Is not a picky eater
- Is easy to catch (this one's important!)
- Is "bombproof," unflappable, and not easily spooked
- Does not bite or kick

Characteristics to Avoid:

- Fights with other horses at pasture
- Displays a bully-type personality
- Requires twitching or sedation to bathe, clip, and deworm
- Is difficult to tie or load
- Chews wood or crib
- Is a nervous eater
- Is hard to catch
- Is easily spooked
- Will bite or kick

A horse should exhibit an adequate amount of bone for the size of his body, and there should be an even ratio of body to leg.

An attractive head is accompanied by a bold eye, well-placed ears, adequate width between the eyes, and large nostrils.

Additionally, you'll also want to consider breed type. Breed associations maintain breed standards that fully describe the ideal characteristics of their breed. These traits are important to understand and keep in mind when searching for a horse of a particular breed.

Care and Shelter

Unless you're planning to board your horse at another farm or at a boarding facility, you'll probably be keeping your horse at your farm, which means

Sometimes the weather will dictate whether your horses stay in the barn or head out to pasture for the day.

What better way to start your day than to be out in the crisp air feeding horses during a glorious sunrise?

you'll be able to be in full control of your horse's daily care and have the ability to oversee its diet and health. Before you make the leap to horse ownership, however, be sure that your property is zoned for horses; some areas have restrictions on the number of horses allowed per acre.

Daily Care

Caring for horses represents a significant commitment. You'll be out to feed them two or three times per day, stalls must be cleaned at least once a day, water buckets must be constantly filled and refreshed, fly spray must be applied, daily exercise must be provided, and vet and farrier appointments must be arranged. Then there are the grooming responsibilities, the daily health checks, and regular maintenance to buildings and fencing.

Horses are grazing animals, and they do best when the bulk of their diet comes in the form of roughage—either grass or hay. The horse's digestive system is designed to have food continually flowing through it, which is why pasture grazing is ideal for many horses. In the absence of this option—due to lack of pasture, wintertime, and so on—you can simulate this experience by feeding multiple small meals throughout the day rather than one or two large meals timed further apart. Frequent feeding is considered to be the healthiest option for most horses.

Ask your veterinarian whether you should also provide grain and/or supplements for your horse. Depending upon your horse's age, condition, and activities, you may need to provide additional calories and nutrients in the form of grain and/or supplements.

Once you've established a feeding schedule, be sure to stick to it. Horses are creatures of habit, and they quickly learn the ins and outs of a particular routine. If they expect you to hand out morning hay at 6 a.m., don't be two hours late just because it's the weekend.

Having access to pasture is a real benefit to the majority of horses. In addition to it being an excellent source of roughage, most horses benefit physically and mentally from grazing. If your horse is obese or insulin resistant and shouldn't have access to grass, you'll want to limit his turnout to a dry lot instead of a grassy field.

If you have a young horse that is still growing and full of energy, then the more hours it has at pasture the better. Easy keeper horses (ones that stay fat on air) usually benefit from having less time out at pasture. Sometimes the weather will dictate how much time you want your horse to spend outdoors. If the weather is bitterly cold and miserable, you might prefer to keep the horse in during the cold nights and early mornings. On the other hand, if it's mid-summer and hot, muggy, and buggy during the day, consider stalling your horse during the heat of the day and turning it out during the cool nights. If your amount of grass pasture is limited, limit the number of hours that your horse has access to it each day so that the pasture quality remains higher than it would if the horse were constantly grazing.

Stalls and Bedding

Providing a stall for your horse is important for many reasons. Extreme weather conditions are a major consideration. If you live in a climate where the weather gets dangerously cold (or dangerously hot), a stall is a very beneficial accommodation for your horse. A stall also allows you easy access to your horse when you decide to ride, and it's a place for the farrier to work or the vet to examine your horse. Many horses come to view their stalls as their "safe haven," and will often be at the paddock gate, waiting for you to put them back inside. Their stalls represent safety, peace, food, and water—all very important things to a horse!

A large, clean box stall with fresh shavings and chew guards installed. Steel bars in front of the windows are an important safety precaution.

Unless your horse is a very large breed, a 12 x 12–foot stall is probably large enough to accommodate your horse, with a 10 x 10–foot stall being acceptable for a smaller animal or pony. Ideally, the stall floor should be dirt, although concrete is a possible solution as long as you provide sufficient bedding. Any stall flooring should be covered with rubber matting—tightly fitted to prevent leakage or heaving—and then covered with your choice of bedding.

Good quality shavings are clean, consistent, and made of pine. Never use walnut shavings because they have been blamed for laminitis in horses. Wood shavings are very absorbent, easy to clean, and fairly cost-effective. Although not as absorbent as other methods, a good straw bed is soft, clean, and when baled properly, it is fairly dust free. Straw of any variety is typically inexpensive, but straw bedding becomes damp very quickly and will take you considerably longer to clean, thus increasing your time spent on stall cleaning and manure removal.

Water buckets should be hung at a comfortable drinking height for the horse without being so low that he could catch a foot in the bucket. Hay racks, if used, should be hung slightly below the horse's eye level to avoid it having to reach up to eat, which can cause it to develop undesirable muscle buildup along the front of his neck.

If the stall is primarily built of wood you will want to install chew guards along the straight edges. Some horses love to chew wood and can get into the habit of it, which if left unchecked can lead to cribbing. All stalls must be checked for safety hazards, such as sharp edges or protruding nails, and the appropriate repairs should be made when necessary. Windows should always be covered with bars or a grid to prevent accidental breakage.

Fencing 101

Several fencing options are available with varying amounts of maintenance and costs to fit a variety of budgets. Board fencing is the traditional form of fencing for horses. It's safe, attractive, and durable when maintained properly. It's also expensive and requires regular attention (for painting and replacing boards, for example). Horses are notorious for chewing on expensive board fences.

Electric fencing isn't as beautiful as wood, but it requires less maintenance.

Many horse owners like to enhance the safety of their board fences by running a single hot wire of electric that is attached to the board around the perimeter of the fence. This increases your horse's respect for the fence, while maintaining the classic look of board fencing.

Although not as aesthetically pleasing as other types of fencing, electric fencing is a popular option for many horse owners. Electric fencing is generally a very safe type of fencing that horses respect. This is a less expensive option than traditional board fencing, as there can be more distance between posts and far less maintenance. Obviously horses will not chew on electric fencing, although they may chew on the posts if they are wooden.

Electric fencing comes in a few varieties, including wire, braided rope, and tape. The safest of these are the rope and tape, as wire is less forgiving in the event of entanglement. Many factors can influence the effectiveness of your electric fence—proper grounding, correct choice of low impedance charger, and the size of the pasture being electrified. You will also need to regularly remove weeds or tall grass that can grow up alongside the fence and short out your electrical current. In addition, you will need to decide if a solar charger will be suitable for your needs or if you will install a higher rated system. Generally, solar chargers work well with small fields that are

Barbed wire is a fence that is best taken down. Any sections of this fence should be removed from your property to protect your horses.

not near any convenient power source. However, these solar chargers can be fairly unreliable, especially in the winter when the sun is less intense or during long periods of cloudy weather.

Vinyl fencing has several advantages. It's less expensive than wood, horses won't chew on it, and there is no maintenance. However, it can be rather flimsy and won't stand up to aggressive use if horses try to reach through or rub on it, so it may not be an effective solution for a pasture situation. These fences sometimes work better for riding arenas where you would like a more decorative fence without the cost of wood. As with wood, vinyl fence can be reinforced with a strand of electric wire to discourage horses from tampering with the fence.

Corral panels are portable, easy to move around, and allow you to set up small paddocks or round pens wherever you desire. Corral panel fences are not as expensive as some of the other types of fences. They can lack durability as horses can (over time) push and bend the panels, rendering them unusable. Depending on the type of finish, the paint can chip off and the panels may rust, but good quality panels retain their finish and do not require painting or additional maintenance.

Barbed wire fencing might have been fine on Uncle Albert's farm when you were a child, but it is definitely discouraged for use as horse fencing nowadays. If a barbed wire fence is already installed on your property, consider removing it and replacing it with a safer form of horse fencing.

Whether your horses are turned out 24/7 or whether they spend some time in stalls, you're going to want some sort of an outdoor shelter in your pasture or paddock area. A three-sided shelter provides a good place for your horses to escape from weather extremes. It offers protection from rain and snow, a windbreak during violent winds, a shady place to escape the hot summer sun, and a place to get away from flies and other annoying bugs.

Your shelter should be large enough to amply house all of the horses in that particular pasture. You want plenty of space for all of them to comfortably rest inside the shelter without the dominant horses keeping the meeker ones from entering.

No hoof, no horse! The services of a farrier will be invaluable to ensure the long-term health, performance, and value of your horse.

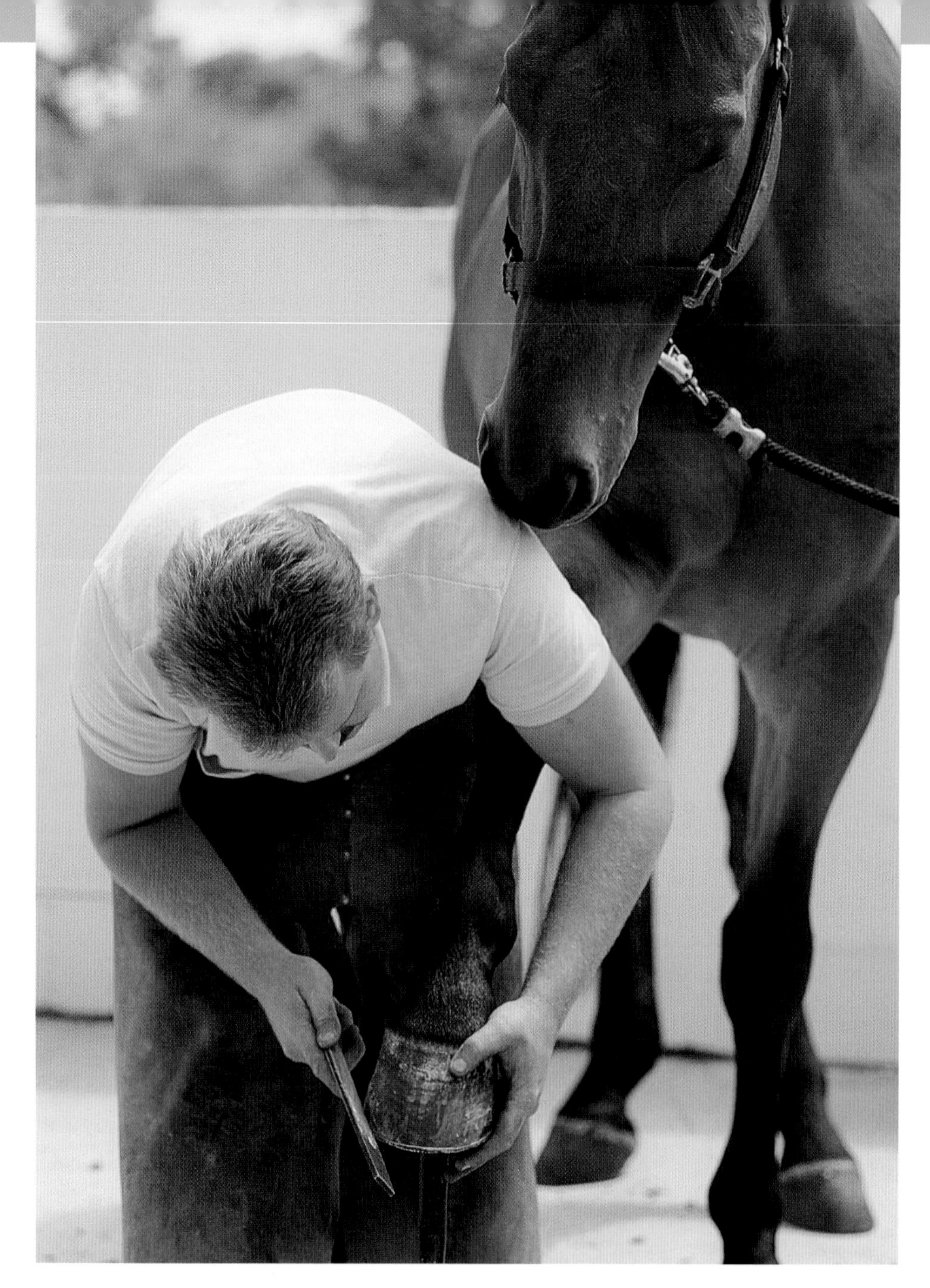

Choosing a Vet and Farrier

These two individuals are an important part of your horse keeping endeavors. A knowledgeable, responsible farrier is highly important, as you're placing the well-being of your horse's hooves in this person's care. Look for a farrier that is experienced, trustworthy, and punctual. Punctuality is more important that you might realize—waiting around for a tardy farrier is a common experience for many horse owners, and one that you'll want to avoid if possible.

Ask for recommendations from other horse owners in your area. They'll be able to point you in the direction of a knowledgeable farrier. The same is true for finding a vet. Ideally, you'll find a vet that specializes exclusively in equines, but you may have to settle for a large animal vet rather than a specialist. Remember, you're trusting your horse's health to this individual, so choose carefully.

RABBITS

If you're planning to raise rabbits, you probably have one of the following goals in mind: raising fancy rabbits for exhibition, raising angora rabbits for their fur, or raising meat rabbits. Or maybe you're interested in the idea of keeping a few rabbits as pets, without the plan of raising any more.

It's important to keep in mind that owning lots of mediocre rabbits is not better than owning a few top quality rabbits. Always aim for quality rather than quantity when raising rabbits. Don't think that you have to run out and buy a dozen rabbits in order to get your rabbitry started. You really only need a trio.

Raising rabbits is a pastime that often interests the entire family. By involving your children in the daily care and management of your rabbitry, you'll be teaching important skills that will be useful in the future.

Buying Rabbits

Your trio of rabbits will consist of one male and two females of the same breed.

Choose your trio with the utmost care, because the rest of your rabbitry will come from them. Ask for advice in selecting the best rabbits. Obviously, health is paramount, so choose rabbits that are free of any contagious diseases of congenital defects. Avoid rabbits with any disqualifying defects, such as malocclusion of the teeth, blindness, mismatched toenails, torn ears, and so on. Do not purchase rabbits with runny noses or rabbits that sneeze, as these can be signs of pastuerella (also known as snuffles), which is a highly contagious respiratory illness that afflicts rabbits.

Study up on breed type and the characteristics that are especially important in your chosen breed. View as many examples of the breed as you

One of the rarest rabbits in North America, the Silver Fox is on the Livestock Conservancy's critical list.

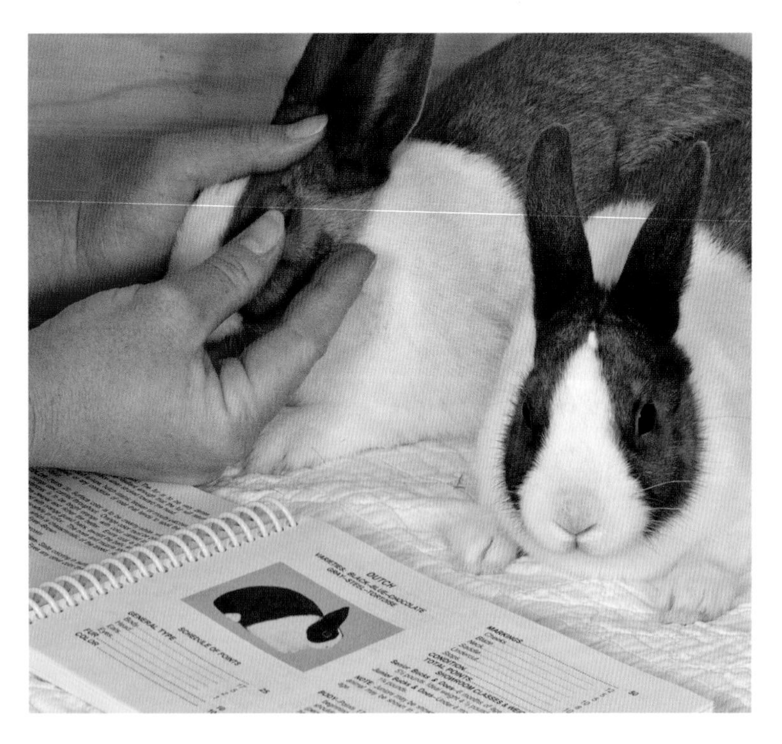

The ARBA's Standard of Perfection should always be your guide when selecting rabbits to breed or show. Evaluate the characteristics as outlined for your breed, and carefully compare your rabbits to the standard.

can, and try to evaluate numerous rabbits against each other and against the breed standard. You will learn to recognize substandard rabbits very quickly.

Rabbits are generally divided into three categories when offered for sale: show rabbits, brood rabbits, and pet quality. A rabbit that is offered for sale as a show rabbit is an exceptional individual that exudes breed type and quality. It is free of any disqualifying characteristics, is the appropriate size for its breed and age, and has the potential to be successfully shown at American Rabbit Breeders Association (ARBA)–sanctioned shows.

A brood rabbit is one that meets the criteria and characteristics of its breed but might not be quite up to the level of quality that is needed for showing. Brood rabbits are sometimes a bit larger than is allowed for showing, rendering them unable to show but perfectly acceptable for breeding purposes, especially when they are mindfully crossed with a smaller rabbit in hopes of producing subsequent generations of the ideal size. Brood rabbits sometimes have cosmetic disqualifications that are not genetic and will not transmit to their offspring but prohibit them from showing.

Pet-quality rabbits are usually rabbits that are not deemed to be of appropriate quality for show or reproduction, and thus they are marketed as pet rabbits to families who wish to add a special bunny to their lives.

Meet the Breeds

There are nearly fifty different rabbit breeds recognized by the ARBA Let's take a minute to go over a handful of breeds.

1. **Holland Lop:** I list this breed first because it's my hands-down favorite—and I'm not alone! Holland Lops have an enormous fan base and are an extremely popular choice for the show ring. Holland Lops weigh up to 4 pounds.
2. **Mini Rex:** The most popular rabbit breed in America—and for good reason. Mini Rex are what many consider to be the perfect size—about 5 pounds—and they have exquisite Rex coats with their velvet-like texture. Mini Rex are noted for their kind dispositions, and they make excellent all-around family rabbits for breeding or show.
3. **Rex:** Weighing in at approximately 8 to 10 pounds, the Rex is larger than the Mini Rex but possesses many of the characteristics that make its miniature cousin so popular.

The Mini Lop is a medium-sized rabbit of compact shape. This one has Blue Opal coloring.

4. **Dutch:** The ultimate child-friendly rabbit! Dutch rabbits are notably calm, laid-back, and sensible. They are found in a handful of attractive colors, all with their trademark pattern of white markings.
5. **New Zealand:** Along with the Californian, New Zealand rabbits are commonly chosen for meat purposes.
6. **English Angora:** If you want to raise rabbits for their wool, you'll need one of the Angora breeds, such as the English Angora. Don't underestimate the commitment that's needed when raising the Angora breeds, however! You'll be taking care of a lot of wool, so be prepared for the journey.
7. **Netherland Dwarf:** On the small end of the scale sits the Netherland Dwarf. With adorable faces and tiny ears, Netherland Dwarf rabbits typify the concept of a dwarf rabbit.
8. **Any breed recognized by the Livestock Conservancy:** Several breeds are currently listed on the conservation priority list that is compiled by the Conservancy. If the idea of saving an endangered breed appeals to you, check out some of the breeds on the list. By helping to increase the population of these rare breeds, you're ensuring that they remain viable for future generations.

Equipment to Purchase

Although getting started with rabbits requires far less equipment and space than you might need for other types of livestock—say cattle or horses—you'll still need several things in order to get your rabbitry set up properly.

At the very least, you'll need the following items for each rabbit:

- Hutch or cage
- Feeder
- Waterer
- Floor mat or resting board

Other equipment:

- Hay rack
- Exercise pen
- Grooming stand
- Carrier (useful for transporting rabbits)
- Fans (to keep your rabbits cool)
- Nest boxes (a must for does and their litters)
- Cage cart
- Scale

Additionally, you'll need a space to keep your rabbits. This area needs to be a safe, sheltered location where they will be protected from heat, sun, adverse weather conditions, and predators. A large garage can be an excellent choice, or a roomy, well-ventilated shed. Some people choose to keep their rabbits outdoors—but you'll need to offer complete protection from hot sun and strong winds, as rabbits are not the best at regulating body temperature and they overheat quickly.

You can purchase rabbit equipment from any manner of rabbit equipment suppliers, or you can shop at your local farm equipment store, such as Tractor Supply Company. You can purchase hutch or cage kits that are easy to assemble and often come complete with plastic trays (to catch droppings and urine) and water bottles, making it easy to get started with a bunny project. Alternately, you can build cages and hutches from scratch,

Inside a small barn is an ideal place to house rabbits, but you can also set up the hutches outdoors. These hutches have solid ceilings to keep out the rain and wire mesh to keep rabbits in and predators out.

LEFT: Although pelleted feed is undeniably important, hay is a truly vital component of a rabbit's diet. Hay contains necessary fiber, and munching on hay also decreases boredom for rabbits.

RIGHT: Wire cage kits snap together quickly and easily, and the wire mesh floors allow droppings to fall through to the pans underneath.

if you're handy with tools and have the know-how for the job. This effort can save you some money on the cage kits but will require more time.

No matter how they are built or assembled, rabbit cages or hutches need to have wire mesh floors that allow droppings and urine to fall through to the ground or to a tray underneath. This construction provides a clean environment for your rabbit and eliminates messes; however, be sure to provide floor mats or resting boards to give your rabbits a comfortable place to rest.

Cleanliness is essential to the health of your rabbitry—so be vigilant in keeping your cages tidy and the trays emptied.

Feeding Your Bunnies

It is true that rabbits are relatively easy to care for when compared to other types of livestock, but it isn't quite as simple as some people make it sound.

Most rabbit breeders feed commercially produced pelleted rabbit food. These carefully formulated products provide balanced nutrition that is hard to match with homemade grain mixtures.

But just as important is hay. It truly is one of the most important parts of your rabbits' diet. Hay provides forage, the opportunity for rabbits to chew, and a beneficial way to prevent gastrointestinal issues.

Access to fresh, clean water must be provided 24/7.

Raising Litters

Gestation in rabbits is approximately 31 days. Kits (baby rabbits) are born blind, deaf, and hairless, but they grow rapidly. The topics of rabbit gestation and kindling (birth) are covered extensively in *How to Raise Rabbits* and *The Rabbit Book*.

A Cochin Frizzle chicken.

A Golden Laced Polish chicken.

A White Leghorn rooster.

A Red Star chicken standing on hay in the barn.

CHICKENS AND OTHER POULTRY

One taste of a home-raised egg is usually enough to convince anyone of the merits of keeping chickens—and if you can raise enough to sell, so much the better. Throw in the potential for organic, free-range meat and the opportunity to interact and enjoy the antics of hens and chicks, and you have the recipe for a perfect farm animal.

Selecting a Breed

The first step is to select your preferred breed of chicken. Here are some tried-and-true classics:

- Plymouth Rock (Barred Rock)
- Rhode Island Red
- Leghorn
- Orpington
- Silver-Laced Wyandotte
- Australorp
- Ameraucana
- Brahma
- Silkie

A free-range hen with chicks.

In addition to these popular breeds, dozens and dozens of less-common and rare chicken breeds are available. More than fifty chicken breeds are listed on the Conservation Priority List of the Livestock Conservancy in varying stages of endangered status. If preserving a heritage breed appeals to you, be sure to check out the list at the Livestock Conservancy website.

Chicken breeds are generally divided into three categories: laying breeds, meat breeds, and dual-purpose breeds that are useful for both eggs and meat.

Additionally, bear in mind that there are bantam versions of many breeds, which are considerably smaller than their large counterparts and require less space to keep. However, they produce fewer eggs than full-size chickens and the eggs are somewhat smaller.

Getting Started

Chickens prefer to be in the company of other chickens, so bear in mind their socialization needs when planning your flock. A single hen by herself is bound to be lonely, so aim for a minimum of three to five chickens for starters.

Many people automatically assume that they will need a rooster in order for their hens to lay eggs, but this isn't the case. If you merely want your hens to lay eggs for your own consumption, the hens can do that all by themselves. If you want to raise chicks, you must have a rooster. Remember, keeping a rooster presents challenges of its own, so only keep one if you need one.

When your chickens arrive, whether from a local source or by mail order, you'll need an appropriate place to house them. For young chicks, this

will be a brooder in which they will be kept warm and dry. If you buy an established flock of older chickens, you'll need a chicken coop.

Chicken coops can be found in a wide range of shapes, sizes, and designs. At its most basic, your chicken coop will provide a sturdy, safe enclosure for your chickens while allowing ample ventilation and appropriate roosting areas and nest boxes for your hens. (You don't necessarily need individual nest boxes for each and every hen; one box for every two or three hens will usually suffice.) Chicken coops that are more elaborate can be impressive structures with wheels to transport the coop from place to place and room for a substantial flock of hens. At a minimum, you'll want to allow 3 square feet of coop space per bird (5 square feet each is better). You'll have to determine the cooping needs of your individual flock and then construct or

CHICKEN LINGO

Need to refresh your memory on chicken terminology?

Rooster: a male chicken over 1 year
Hen: a female chicken over 1 year
Chick: a baby chicken
Pullet: a female chicken under 1 year
Bantam: a small breed of chicken
Cockerel: a male chicken under 1 year

EGGS OF DIFFERENT COLORS

White eggs are wonderful in their own right, but colored eggs can add a rainbow of delight to your egg-gathering duties! Try one or more of these color-producing breeds and add some excitement to your eggs! (Keep in mind that the precise shade of egg color can vary depending upon the individual hen.)

Blue Eggs

- Araucana
- Ameraucana
- Cream Legbar
- Easter Egger

Pinkish Eggs

- Easter Egger
- Dorking (occasionally)
- Delaware (occasionally

Brown Eggs

- Australorp
- Jersey Giant
- New Hampshire Red
- Wyandotte
- Cochin
- Rhode Island Red
- Plymouth Rock (can be pinkish)
- Brahma
- Dominique
- Java

Dark Brown/ Reddish Eggs

- Maran
- Barnevelder
- Penedesenca
- Welsummer

Green Eggs

- Easter Egger
- Ameraucana

Egg cartons with a farm's label. These will be sold at the farmers' market.

A cute sign can draw shoppers to your fresh eggs.

purchase an appropriate enclosure to suit those needs.

In addition to your coop, you'll want to construct an outdoor run for your chickens. Aim for an area of at least 10 square feet per bird, and even more is better. Although chicken wire sounds like it would be the appropriate choice for enclosing chickens, hardware cloth—with its tiny openings—is a better choice to deter predators.

Once you've determined your cooping arrangements, you'll want to select feeders and waterers.

Selling Eggs

Let's say that your young pullets have been growing steadily and are now beginning to produce eggs on a daily basis. Generally speaking, you can figure on one egg per chicken per day, but this is assuming that your chickens have at least fourteen hours of light per day, as this is necessary for egg production.

Follow food safety guidelines for the proper handling of your eggs in order to maintain their quality. Aim to gather eggs at least twice per day (or even three times), then clean the eggs properly, and keep them in a refrigerator.

You've seen the stereotypical eggs that are sold in supermarkets—they're white (sometimes brown), uniform, and flavorless. If you'd like to pursue an egg-selling endeavor, try to increase your profitability by raising and selling eggs that exhibit the following characteristics:

- **Free-range:** Consumers like free-range eggs, gathered from chickens that are allowed to roam about in grassy areas.
- **Organic:** Eggs that are produced without pesticides, synthetic hormones, and antibiotics are very popular with consumers.
- **Colorful:** You can often command a higher price per dozen for colored eggs at farmers' markets.

When preparing to sell eggs, either at a farmers' market or at your farm, be sure to investigate any state or local regulations pertaining to the sale of eggs in your area. Some localities require licensing and/or inspections for farms that sell eggs.

OTHER POULTRY

Ducks and Geese: These waterfowl can be lots of fun to have around the farm, but bear in mind that they do benefit from the presence of a small pond. Ducks and geese tend to be a bit messier than chickens, but they are endearing creatures.

A Silver Appleyard duck hen with her ducklings.

ABOVE: Four beautiful Toulouse geese.

RIGHT: A free-range Royal Palm turkey.

Turkeys: Turkeys can also been an intriguing addition to your rural menagerie. Larger than chickens, they produce more meat, and there's a strong (and growing!) market for heritage turkeys.

Ostriches: The flightless ostrich is the largest bird in the world and is the only two-toed bird. Its lifespan is lengthy—thirty to seventy years—and

LEFT: Muscovy ducks are unmistakeable.

RIGHT: An emu in its pen.

farmers raise ostriches for their meat, leather, feathers, and eggs (which can weigh between 3 and 5 pounds).

Emu: Like the ostrich, the emu is a member of the ratite family and is the second-largest bird in the world. Emu are raised for meat, leather, feathers, and eggs, but they are also well known for their oil. Emu eggs are a brilliant emerald green and their meat is low in cholesterol and fat while being high in iron and protein.

Peafowl: Peafowl come in two main types—Indian and Green, with the former considered to be hardier and the latter considered to be less cold tolerant. Peafowl typically weigh between 6 and 13 pounds, and although they can fly, peafowl frequently prefer to stay on the ground. They do sometimes roost in trees at night. Their average life span is twenty to forty years.

Quail: Like chickens, turkeys, and peafowl, quail are *galliformes*. Quail are often raised for meat and eggs, and there are two common varieties: the bobwhite and the Japanese (also known as Coturnix) quail. The bobwhite is characterized by its distinctive "Bob-White" call and its white eggs; Japanese quail produce speckled eggs. The average life span for quail is two to five years.

BEES

Bees make honey and beeswax, and they provide pollination of crops, vegetables, and flowers (bees are said to be responsible for 80 percent of insect crop pollination in the United States). Bees are fascinating creatures that have unbelievable organizational skills and social habits. To produce a single pound of honey, worker bees must visit as many as two million flowers while traveling an accumulated distance of fifty thousand miles.

Bees do need regular care, but not the same level of day-to-day commitment that livestock such as cattle or horses require. Bees work hard, take care of themselves (for the most part), and provide an incredible return of honey and beeswax for minimal investment of your time.

Getting Started with Honeybees

Of course, you already know that you'll need bees, but what else do you need to get started in a beekeeping venture? Here's a quick checklist of items:

A package of bees typically consists of about three pounds of bees, including a queen and about ten thousand workers. The bees travel with a can of syrup to feed them during their journey.

Did you know that 80 percent of crops are pollinated by honey bees? Some beekeepers travel the country with their hives and bees, offering pollination services to farmers and orchardists.

- Protective clothing (a full body coverall, gloves, and a veil)
- Bee smoker (a necessary tool for working with bees; you'll use it whenever you inspect your hives)
- Hive tool (useful for opening hives and manipulating frames)
- Bee brush (useful for brushing bees out of the way)
- Frame grip (a clamp-like tool that is used to pull frames from the hive box)
- Spray bottle (you'll fill this with sugar syrup when you install your bees in their new hive)
- Feeders (entrance feeders, pail feeders, hive-top feeders)
- Hives, which can include the following components: a hive stand, a landing board, a bottom board, hive boxes, supers, an inner cover, an outer cover, and an entrance reducer.

Additionally, you'll need frames to fit in your hive boxes and supers, and you may want to consider other items, such as a queen excluder.

Once you've amassed the appropriate equipment (available from any number of beekeeping supply companies), it's time to purchase your bees. Now you might be asking yourself a very important question: Just where are these bees going to come from?

Since it's unlikely that a swarm of homeless honeybees will buzz their way into your life (and as a beginner, you probably couldn't catch them even if you wanted to), you'll have to purchase your bees from a bee breeder or a local beekeeper. Many bee breeders in the United States are located in the southern portions of the country and in California, thanks to the warm temperatures.

MEET THE BEES

In the hive, there is an impressive cast of some fifty thousand characters, but they can be generally grouped into three categories:

Queen: This gal is super-special; she is the only egg-laying female in the hive. Because of this, the other bees in the hive provide her with the best care and treatment. Queen bees are much larger than the worker bees.

Worker bees: These bees make up the bulk of the colony. These hardworking individuals perform an entire host of jobs in the hive: from tending the queen, tending the nursery of young bees, building comb, storing food, and flying from flower to flower colleting pollen and nectar.

Drones: Male bees are called drones, and they are notable for their large eyes and their lack of contribution to the hive. Other than their mating flights with the queen (which is undoubtedly helpful to the hive), the drones do not work.

LEFT: Essential beekeeping equipment: a full body coverall, gloves, a veil, a smoker, and hive equipment.

RIGHT: The Langstroth hive design, developed in the nineteenth century, revolutionized beekeeping and is still the most popular hive design today. Other hive types include the top bar hive and the Warré hive.

Luckily for you, these bee breeders ship packages of bees all over the country. In all likelihood, you'll either purchase your bees directly from one of these breeders, or you'll purchase from a local supplier who brings in large numbers of packaged bees and distributes them to area beekeepers. In either case, you'll probably be purchasing what's known as a "package" of bees. This package consists of a small screened box that contains approximately 3 pounds of worker bees (which equates to roughly ten thousand bees) and a queen. You will then install these bees directly into your waiting hive. (This is a complex process with many precise steps. See *The Beginner's Guide to Beekeeping* for step-by-step instructions and photos of the process.)

Once your bees are installed into their hive (usually in spring), they'll begin the fascinating-yet-labor-intensive process of working the flowers and blossoms to gather pollen and nectar and begin producing honey and wax.

You'll want to inspect your bees periodically throughout the summer to ensure that everything is running smoothly in the hive. Your routine inspections will involve removing the outer cover of the hive and exploring the frames inside your hive boxes or supers. Look to see if the queen is laying eggs, check if the eggs are developing into larvae, and make sure that the bees are drawing out comb and they are storing honey. These steps generally

After you install your bees into their hive, they will begin working any flowers and blossoms that happen to be in bloom at that time.

occur in a natural progression without much outside assistance from you, but it's important to keep an eye on the hive because you can ward off potential problems by paying attention. If you discovered that your hive is empty of eggs and larvae, it would be a good indication that your hive is queenless and you may need to introduce a new queen. Or if you noticed that the worker bees are building queen cells, this may be an indication that the hive is in danger of swarming. By watching for subtle signs in your hive, you can promptly attend to any situation that arises.

The Honey Harvest

It's important to remember that you must wait to harvest honey until your bees have established a substantial supply. This step is vital because you'll want to ensure that you leave behind a suitable amount for the bees' use during the winter months, since they must have enough left over after you've harvested. A good rule of thumb is: bees first, then you.

But let's say that your hives are overflowing with an abundance of golden glory and there's plenty for the bees to share with you. In this case, the prime object is to remove the honey from the hive.

To begin with, you'll need to gather the necessary equipment with which to harvest your honey. For starters, here's what you'll need:

- **Extractor:** This is the most important piece of honey harvesting equipment. It's a large, stainless steel drum that is either hand-cranked or motorized, and uses centrifugal force to extract the honey from the frames as the frames spin around in the drum.

It's important to perform regular, routine inspections of your hives during the summer months. You can potentially avoid problems by keeping a close watch on your bees.

BELOW LEFT: You'll need to invest in honey extracting equipment, including an extractor.

BELOW RIGHT: Frames of uncapped honey are placed inside of the honey extractor, where the spinning—the centrifugal force—extracts the honey from the frames and cells.

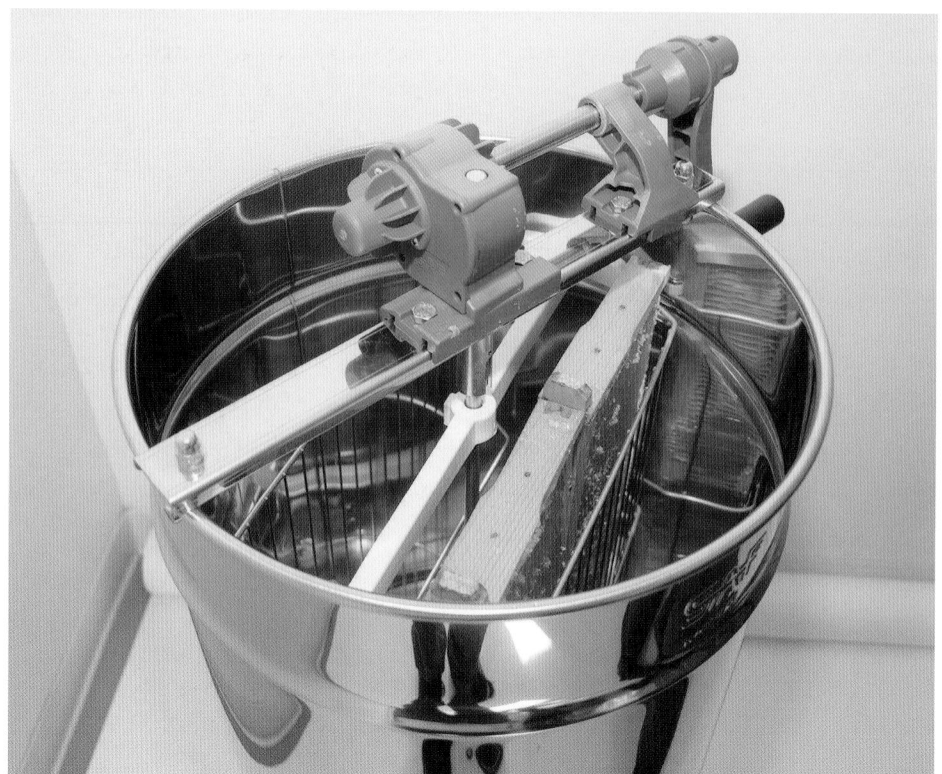

If you're planning to market your honey and other products from the hive, be sure to aim for high-quality products that are nicely packaged and attractive.

- **Uncapping tank or tub:** This piece of equipment is where you uncap your frames of honey. The loose chunks of wax and honey are collected in this tank during the process.
- **Uncapping knife and fork:** These tools are handy for the uncapping process and are budget friendly. (Note, some uncapping knives are heated, which can make the uncapping process easier, but it increases the cost of the item.)
- **Strainer:** Believe me; finding debris and random bug parts in your gorgeous honey is highly unappetizing. Strain those bits and chunks out using a strainer.
- **5-gallon container**: But not just any 5-gallon container. Your extracted honey will be collected in this container, so be sure to obtain it from a beekeeping supply company and make sure that it's made of food-grade plastic. Five-gallon containers for honey extraction are equipped with a closeable gate at the bottom which makes bottling the honey a snap.
- **Bottles:** You'll want to purchase an ample supply of bottles in which to store your honey; these bottles come in a range of sizes (1-pound bottles are popular), shapes (bear-shaped, anyone?), and can be found in plastic or glass. Choose the bottle type that best meets your needs or experiment with a few different types.

The process of extracting honey requires several steps, but essentially begins when you remove the frames of capped honey from your hives and ends when you place the caps on those golden bottles of honey. As you ponder the joy of the honey harvest, don't forget the hours upon hours of work (and the thousands of miles of flying!) that your bees put into the honey. Their incredible efforts are truly remarkable.

Don't forget about your beeswax! Your leftover cappings can be transformed into candles and other products.

Expanding Your Horizons

When the busy summer season is over, you might be wondering, what next? Well, if you're fortunate, you'll be able to overwinter your bees and keep them alive during the winter months. In warm climates, this is easily done, but in cold, northern climates, it's much harder to overwinter. You'll want to provide your hives with protection from the elements and plenty of nourishment. Your contributions can be in the form of stored honey that your bees have saved from the summer months, or you can provide candy boards to keep your bees happy and nourished during the winter.

If you expand your beekeeping operation to include multiple hives, then it's natural to begin thinking about the possibility of marketing your extra honey, beeswax, and pollen to local consumers. Before you set up that farmers' market stand, be sure to explore any state and local ordinances and regulations that might apply. Some states require you to obtain a food

BUT WHAT ABOUT BEESWAX?

Don't overlook this additional treasure from the hive! After you've extracted honey, you'll be left with a pile of cappings that were removed from the frames during the extraction process. If these loose cappings are rendered (melted down and filtered), you'll be left with beeswax that is suitable for a range of projects, including the following:

Beeswax candles: These are easy to make and make fabulous gifts. Any beekeeper should make beeswax candles at least once.

Cosmetic items: Thanks to its anti-inflammatory and antibacterial properties, not to mention its germicidal antioxidants, beeswax is a useful product for cosmetics such as face cream, moisturizer, and hand salve.

Lip balm: Like candles, beeswax lip balm is super-easy to make and fun, too!

Furniture polish: Shine up those tables with homemade furniture polish!

Modeling wax or crayons: Share your hive bounty with the children in your life—who doesn't love a colorful crayon or modeling wax?

producer's license or a food handler's permit before you can sell your honey or honey products, so investigate these regulations carefully.

By the same token, there may be regulations and ordinances in your area that pertain to the keeping of bees in general, so investigate all of this before purchasing your first bees and equipment. Some states require beekeepers to file for an annual license; other states require annual inspections by an apiary inspector. Some cities may prohibit beekeeping entirely or enact severe regulations that make beekeeping virtually impossible. Rural areas usually don't have a lot of restrictions on beekeeping, but you'll still want to check things out before getting started.

ANIMAL HEALTH

If you keep livestock of any kind, you'll need to learn how to keep your animals healthy. Start off right by purchasing only healthy stock. Learn the basics of birthing and how to care for newborn animals. Teach yourself to recognize the signs of illness, and prepare yourself for handling an animal's death.

Learn how to recognize a healthy animal before you start buying livestock for your farm.

Assessing Health at Purchase

You should be concerned about three important considerations in every animal purchase regardless of species: condition, health, and mobility.

Condition is the term used that refers to the amount of weight or fat the animal carries on its body. Animals that will be pastured should not have excess fat because they will not gain weight very well during the first month as they make the transition to your farm. Animals should be healthy and fit looking. If possible, avoid animals that have not performed well elsewhere. They may be less expensive, but they won't achieve an acceptable performance for you either.

Health is the animal's general physical condition. Animals you purchase should have bright eyes and hair or fleece coat and not a dull, emaciated look. Normal breathing should be observed. Avoid those with runny noses or the lack of a bright eye. Try to locate any ear vaccination tags as this will help indicate the care they have received from the time they were young. Any farmer who pays attention to having his or her animals vaccinated for brucellosis or other viruses and is willing to pay the costs involved is usually concerned with the overall health of the farm's animals.

Mobility is another health consideration. Simply watch them move around. An animal should have the ability to move about freely with no leg, joint, or feet problems. Avoid animals that limp and have swollen joints, long hooves, or other physical impairments. Animals exhibiting any of these conditions do not last long on any farm.

Keeping Your Animals Healthy

One rule that can define any farmer's experience is that animals can get sick even with the best of care. The key is to minimize the number and severity of those illnesses to the greatest extent possible.

The first step in minimizing the effects of any sickness is to visually inspect the animals on a daily basis. This does not necessarily require a great amount of time, depending on your herd size, but walking out to where the animals are and looking for signs of any health or physical problems will help keep you abreast of their condition. You will be able to quickly identify animals that may be listless, limping, have droopy ears, or exhibit other signs that they may not be eating normally to indicate some problem.

Identifying such problems will allow you to move quickly to assist them. Generally, an animal that exhibits symptoms of some illness will not overcome it without your intervention or help. You become their diagnostician, doctor, veterinarian, and possibly, pharmacist all at once. You can treat most illnesses without any specialized education. Experience is the best teacher, but you can gain insight and understanding by reading about your chosen species' health issues and the practical application of administering treatments.

There will be times when you have no choice but to call a veterinarian for help. However, the severity and extent of the problem may have to be solved by

An alert horse with a gleaming eye and shining coat is generally one in excellent overall health and condition.

your own initiative, such as a case of bloat, where time is extremely short and a call to a veterinarian most likely will be too late. A licensed veterinarian is required to administer vaccination injections, medical euthanasia, and certain antibiotics or steroids.

Conventional and Alternative Treatment

Conventional treatment of livestock generally involves the use of antibiotics to relieve the symptoms of the illness or disease. Antibiotics may correct the problem in the short run, but residual effects may affect marketing the treated animals.

The use of antibiotics to treat disease in food-producing animals started in the mid-1940s, and by the early 1950s, the industry saw the introduction of antibiotics in commercial feed for livestock. In the past forty years, antibiotics have served three purposes: as a therapy to treat an identified illness; as a prophylaxis to prevent illness in advance; and as a performance enhancement to increase feed conversion, growth rate, or yield.

Bacterial diseases generally cause pain and distress in the animal, as well as an economic loss if unchecked. Antibiotics can be used to help reduce this suffering and distress and speed the recovery of an infected animal.

When used responsibly, antibiotics can be an essential element in the fight against animal diseases. In rare instances, they may be used to prevent diseases that might occur in a herd or group of animals if a high probability exists of most or all the animals becoming infected.

However, there are some cautions when using antibiotics, especially on a routine basis. In animals, as in humans, a significant proportion of those treated for infectious disease would recover without antibiotics. One study has shown that every year, 25 million pounds of antibiotics, roughly 70 percent of the total antibiotic production in the United States, are fed to chickens, pigs, sheep, goats, and cattle for nontherapeutic purposes, such as growth promotion. This report also showed that the quantities of antibiotics used in animal agriculture dwarf those used in human medicine. Nontherapeutic livestock use accounts for eight times more antibiotics than human medicine, which uses about three million pounds per year.

Any animal that has received antibiotic treatment must be withheld from the market for a specified time, and the withholding times must be followed by the producer or veterinarian. This regulation doesn't mean that antibiotics shouldn't be used. But if they are used, it is imperative that they be used judiciously and only when warranted. In some specific cases, such as Johne's disease, it is better for the animals to be culled rather than put them through any treatment program because in almost all cases, continuing their existence is ineffective and uneconomical.

Familiarizing yourself with diseases that require antibiotic assistance will help avoid unnecessary usage. A discussion with your local veterinarian may provide other insights that will be useful in your program.

Interest has been growing in using alternative treatment programs for animals because of the concern of residual effects of antibiotics, both in animals and the human food supply chain. Information is available for homeopathic treatment systems, and this approach is ideal for those considering organic, sustainable, or biological farming methods.

Homeopathy and herbal treatments have a place within any farm health protocol. One of the ideas homeopathy is based upon is that bacteria are not necessarily a bad thing and that they do not need to be destroyed. It is not the illness that is being treated but the animal's reaction to it.

Every day, walk out to where your animals are and check for problems. This will help you catch signs of illness early on.

Homeopathy treatment involves the natural stimulation of the animal's immune system so that it can fight off the bacteria that might otherwise cause a disease. It is known that antibiotics have a suppressing effect on the animal's immune system, because they indiscriminately fight both the good and bad bacteria.

Providing the animal with the ability to use its own body to fight off disease-causing bacteria benefits the animal in the long run because its physiological system is in better condition. Building up the health of the whole herd or flock and increasing the resistance of its individual members to disease will help induce greater growth rates and overall health. Homeopathy is a legitimate route for treatment of animals but is not a substitute for preventive measures like good nutrition, air quality, and proper sanitation.

A second alternative treatment is herbal, which also uses remedies based on preparations made from a single plant or a range of plants. Applications are by different routes and methods depending on the perceived cause of the disease condition.

These applications can be made from infusions, powders, pastes, and juices from fresh plant material. Topical applications can be used for skin conditions, powders that can be rubbed into incisions, oral drenches for treatment of systemic conditions, and drops to treat eyes and ears. Information is available from alternative stores or books published on these topics to help explain more about these products and their usefulness.

Common Health Problems

Many health issues have the potential to affect the performance of your animals and their lives. Good planning and management, along with use of common vaccines, will usually enable your livestock to avoid most disease problems.

Discussing these problems is not meant to frighten you from pursuing a program of raising animals of your choice; it is meant to alert you to their existence. These are only a few of the health issues that can have possible impacts on your livestock. Generally speaking, these problems do not exist on all farms and may not be a cause of great concern.

Pastured animals appear to require less veterinary assistance and exhibit fewer illnesses because they live in the open air unencumbered by effects of unnatural housing that allows their healthy bodies to ward off infections. You should learn as much as you can about potential problems and have an understanding of how to react when they do arise. This is good husbandry.

It is wise to find a local veterinarian to consult about a health program. A list of veterinarians in your area or region is usually available from your county agriculture extension office and should be consulted prior to bringing any animals onto your farm.

Some common health problems can be handled on your own without the assistance of a veterinarian, including bloat, scours or diarrhea, pneumonia, clostridial diseases, parasites, and infections of the skin.

Bloat is a condition where immediate attention and action is required. Ruminants such as cattle, sheep, and goats spend many hours a day chewing their cud and belching. If that process is severely hindered, the gases built up in the rumen are unable to escape.

As the gas pressure increases, the cow stops chewing and the rumen will swell. In the meantime all the internal organs—lungs, heart, liver, and intestines—are being squeezed by the enlarged and tightly pressured rumen, which makes them work harder. In some cases the death of the animal is due to a heart attack, but most die from asphyxiation because the pressure has squeezed all the air from the lungs.

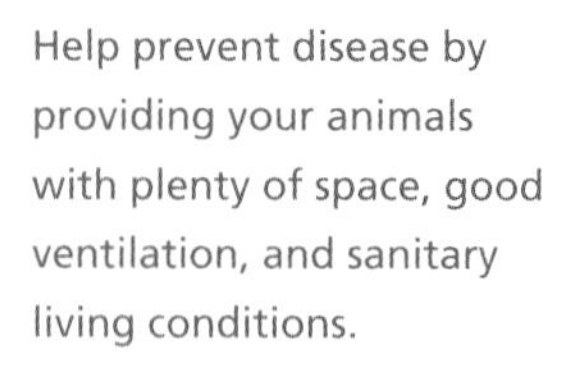

Help prevent disease by providing your animals with plenty of space, good ventilation, and sanitary living conditions.

Animals that have access to pasture tend to be healthier and require less attention from the vet.

Diarrhea (scours) is a common aliment of newborns. The prevention of diarrhea is often easier and more successful than treatment. When animals develop diarrhea, dehydration is a prime cause of death. Electrolytes and fluid given orally or intravenously every two hours is the most effective method of treatment.

Pneumonia usually occurs due to stress, changes in weather, and infectious agents all being involved at similar times. Pneumonia is most common in young that have been weaned or kept in wet bedding conditions and are subject to drafts. Keeping animals in a dry, draft-free environment will help prevent pneumonia.

Clostridial diseases are a group of related infections that may cause sudden death, especially in young animals. These diseases include blackleg, enterotoxemia, malignant edema, and black disease. Vaccines are available to stimulate the immune system of the animal. Cattle should be vaccinated early in life with appropriate boosters given at later dates if these diseases are a problem in the area where you farm.

Internal parasites may be a problem when animals are grazed on the same pastures several years in a row. Deworming may be needed to minimize the parasite load in the body of the animal to allow proper rate gains. Products are available for use at specific times during the year and your local veterinarian can provide you with advice on how best to approach any problem. External parasites, including lice and horn flies, can be controlled with powders and other products readily available.

Brucellosis is a bacterial infection that causes abortion in pregnant animals of any species. It primarily affects the females, but males can also contract this infection. The infected females exhibit symptoms that may include abortion during the last trimester of pregnancy, retained afterbirth, and weak calves at birth. Infected females usually abort only once and develop a level of immunity to further infections.

Brucellosis is a reportable disease in any species, and a single suspected case will cause the herd or flock to be immediately quarantined. This measure may seem extreme, but because brucellosis is highly contagious for other animals, it will spread quickly, although not all animals carrying the infection may exhibit symptoms.

You should avoid purchasing animals that do not have visible signs of being vaccinated, such as state-issued, colored ear tags. The best ways to avoid a herd infection are to have your animals vaccinated at the appropriate age and to buy those that come from herds or flocks that have maintained a routine vaccination program.

Pinkeye is the common name for the reddening of the animal's eye caused by bacteria. It is highly contagious, and if left untreated, it can affect the eyes and possibly render the animal blind in the eye. Typically this condition occurs during the warm months and appears to be transmitted from one animal to another by flies that congregate around their faces. Pinkeye can be treated with powders, ointments, applications of antibiotics, or with the use of homeopathic remedies. If you treat an animal with

Because sheep graze low to the ground, they are more likely to have problems with internal parasites.

pinkeye be sure to use good hygiene during and after treatment to avoid contaminating yourself. Humans can get pinkeye from animals.

Ringworm is a contagious infection of the skin and is easily recognized by the circular white encrusted spots on the skin. The infection primarily affects young animals and is caused by a fungus, which is more difficult to treat and takes longer to eliminate from the animal. One of the oddities of this fungus is that it may lie dormant for several years and seems to be gone until conditions become right and your animals experience another outbreak.

The fungus that causes ringworm can be killed by disinfectants, such as iodine and glycerin that penetrate the scabs and saturate the spots. You can prevent outbreaks of ringworm by not bringing infected animals into your herd. The infection can be largely controlled by maintaining clean surroundings and ensuring good ventilation in barns where animals are kept.

Because ringworm is highly contagious and easily transferred from animals to humans, precautions must be taken to ensure that you or members of your family do not come in contact with infected areas. Using disposable elastic gloves when treating the infection and cleaning out a pen where infected animals live will help prevent the spread of infection.

Warts are caused by four different types of the virus *papillomavirus* and usually affect young animals more than adults, although adults are not

MAJOR ANIMAL DISEASES

Beef/Dairy/Bison:	Swine:	Goats:	Sheep:
Environmental	Atrophic rhinitis	Blue tongue	Blue tongue
Brucellosis	Brucellosis	Brucellosis	Brucellosis
Bloat	Hog cholera	Caprine arthritis encephalitis	Coccidiosis
Johne's Disease	Paravirus	Caseous lymphadentis	Enterotoxemia
Clostridial Diseases	Pseudorabies	Chlamydiosis	Ovine progressive pneumonia
Diarrhea (scours)	Porcine Reproductive and Respiratory Syndrome (PRRS)	Diarrhea	Pneumonia
Footrot	Swine erysipelas	Johne's Disease	Scrapie
Mastitis	Swine dysentery	Toxoplasmosis	White muscle disease
Pneumonia	Transmittable gastroenteritis	Urinary calculi	
Pinkeye		Vibriosis	
Ringworm			
Warts			

As an owner of livestock, you'll need to learn how to administer medicine and other remedies.

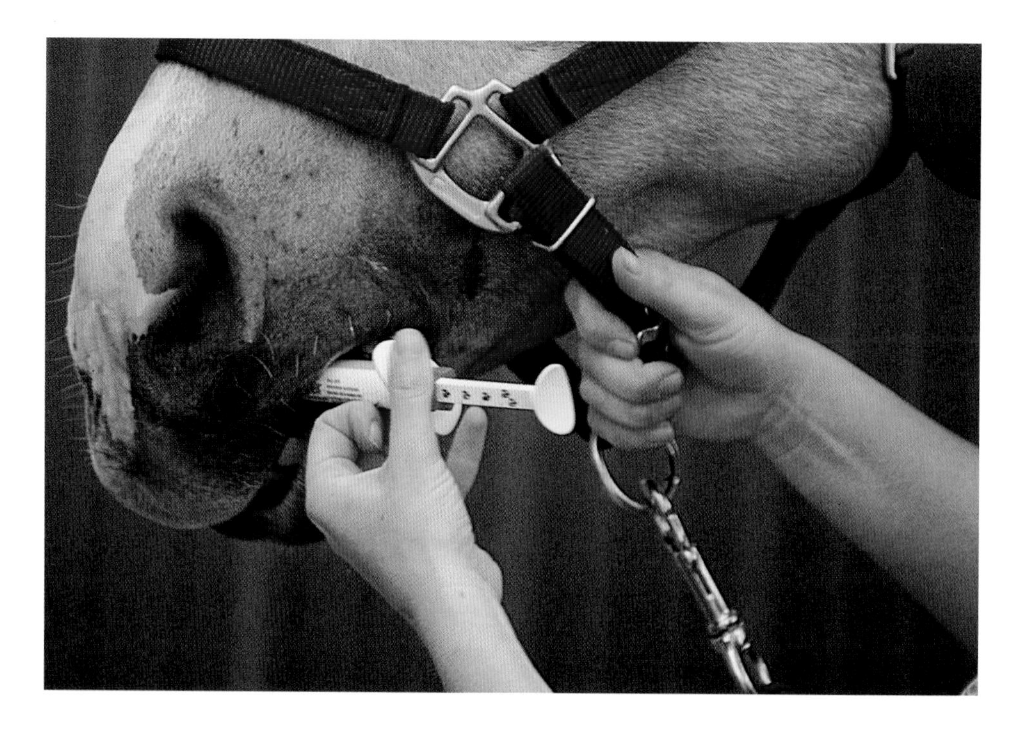

immune from contracting it. Warts are usually more of an appearance problem than a physical problem. Their sudden appearance usually coincides with a sudden disappearance several months later. Vaccines are available, but most warts will disappear on their own if they are left alone.

Euthanasia and Death

Death loss on a farm is inevitable, and producers must be prepared to properly dispose of their dead stock, whether the death came naturally or through euthanasia. There comes a time on every farm when an animal suffers an injury, illness, or some other debilitating condition that requires euthanasia to provide a swift and humane death, quickly relieving its pain and suffering. This should be done in a manner that will minimize any stress and anxiety experienced by the animal prior to unconsciousness.

Correct euthanasia procedures produce rapid unconsciousness, followed by cardiac arrest and finally, total loss of brain function. Several methods can be used to euthanize an animal including chemical, which is always administered by a licensed veterinarian using a barbiturate product. Animals euthanized in this manner should not be used for human consumption. After all vital signs of a euthanized animal disappear, the body must be disposed of.

A physical method of euthanasia that does not require human contact with the animal is gunshot. Strict firearm safety must be observed as well as local laws and ordinances relating to the discharge of a firearm.

Disposal of a fallen animal is a biosecurity issue that is becoming more important and more difficult to handle. Dead animals can be disposed of in several acceptable ways, and the one that fits your farm is an individual decision. Most states have regulations relating to the disposal of dead animals.

Burials are allowed in most states, but you must conform to certain depths of burial; time limits; and distances from wells, adjoining properties, waterways or streams, lakes, and residences to name a few. About three hundred licensed rendering facilities operate in North America, but because geographical voids exist, it may not be possible to get service from one of them.

Information on guidance and assistance as well as ordinances in your area or state can be obtained from your state department of natural resources or your county agricultural extension office

Donkeys are intelligent and fierce protectors of the herd.

Livestock Guardians

Livestock guardians are animals that can display aggressive behavior as a deterrent toward intruding predators and may include guard dogs, llamas, donkeys, and cattle. The choice depends on the size of your flock or herd, your location, the size of your farm, predator threats, budget, and personal preference. Depending on the animal you choose, it will need to be cared for in a manner that keeps it healthy and alert to any potential threats. This may include considerations for feed, training, veterinarian care, and housing costs.

There are two types of sheep dogs—guardian dogs and herding dogs—and each has different behavior patterns and objectives. Guardian dogs are used to protect sheep and goats from predators; herding dogs are used to manage them. Livestock guardian dogs are often referred to as sheep dogs since they most often have guarded flocks of sheep; however, they are capable of guarding other species of livestock as well. Guardian dogs originated in Europe and Asia where they had been used to protect livestock from predators for centuries. The most widely known breeds of guardian dogs in the United States are the Great Pyrenees, Anatolian Shepherds, Akbash, and Maremma.

Guardian dogs are generally large, weighing 75 to 100 pounds, and protective but still can be gentle, make good companion dogs, and are often protective toward children. With the right socialization and training, they can be used as family pets and home protectors. Because of their size, they are generally intimidating and typically remain aloof toward strangers. They typically don't reach maturity until about three years of age.

The purpose of a guardian dog is to bond with the flock and not the shepherd or owner so that their protective instinct is for the animals and little else. They must be courageous in the face of a predator, and they must accept the responsibility of their job to be effective. Training for a guardian dog begins as a puppy, when it should be placed with lambs to develop an imprint on the sheep. This imprinting process is thought to be largely through the sense of smell and happens when the puppy is between three

Llamas can be herd guardians for sheep or goats.

The Maremma guard dog is one of the best known breeds in the United States.

and sixteen weeks of age. They tend to be species specific, as a guardian pup raised with sheep will not be effective with goats or cattle or the other way around. This imprinting appears to be critical and should be accompanied with appropriate training. A good guardian dog will never hurt the sheep it is protecting. They live with the flock they are guarding and bed down with the sheep.

As a guarding dog matures, it spends much of its time near the sheep and keeps watch for outside animals that may approach the flock. Often, if it chases an intruder away, it will return to the flock. Guardian dogs that follow the flock will be on hand if a predator attacks. A successfully trained guardian dog will find that the flock will surround him or her during the period of a perceived threat and will stay close until that threat dissipates.

Herding dogs differ from guardian dogs in their purpose. Herding dogs are not typically used for guarding the flock, although they may serve that purpose if properly trained. Herding dogs, such as border collies, are used to move the flock from one position to another. Their training is different from guardian dogs and they often cannot work both requirements.

Llamas have been used successfully as guardians because of their natural aggression toward coyotes and dogs. A tall, alert, 300-pound llama can

When cattle are in the same pasture as sheep or goats, there tend to be fewer predator losses. But you'll need to raise the animals together, so that they bond, otherwise they'll graze in separate areas of the pasture.

be intimidating to a coyote. Because they are ruminants, llamas can eat the same diet as the sheep they are guarding.

Donkeys have been used for centuries to guard sheep and other herding animals. Donkeys are extremely intelligent and have acute hearing and sight to detect any intruder. They are protective of the flock and, because their size, tower over most predators, which make them effective deterrents. They also eat the same forages sheep eat, so no special supplementary feed is needed to keep donkeys on your farm. Only a gelded jack or jenny should be used as a guardian since intact males can be aggressive toward other livestock.

An often overlooked livestock guardian is cattle. Placing cattle together in the same pasture as goats or sheep tends to reduce predator losses. However, these different species must be encouraged to bond together, otherwise they will tend to graze in separate areas of the pastures and the effectiveness of the cattle guardianship is lost. Young lambs and kids can bond with cattle by initially penning them near to each other in close confinement. When they are placed on pasture together the lambs and kids often will follow the cattle. When a perceived threat from a predator appears, the kids and lambs will stay close to the cattle.

Chapter 4
Repair and Maintain

FENCES

Adapted with permission from *How to Build & Repair Fences & Gates* by Rick Kubik (Voyageur Press, 2007; revised and updated as *Farm Fences and Gates*, 2014)

Fences and enclosures are one of the most important considerations when raising livestock and growing crops because they not only mark the property lines but also humanely enclose your animals and keep out pests. For generations, fences have marked boundaries and given the owners protection. The main purpose of your perimeter fences is to keep your animals in and your neighbor's animals out. They can also be used as barriers against predators.

Design your fence to be highly visible to animals, and consider both strength and aesthetics.

If you already own a farm, it is likely you know the property lines or perimeter of your land. If you are considering purchasing a farm, you'll need to fully understand where the property lines are located. Knowing the exact boundaries of your farm is good business and will protect you and help avoid problems should questions arise if and when adjoining farms are sold.

Perimeter Lines

Before you sign an agreement to purchase a farm, ask to walk and view the perimeter or fence lines with the owner or real estate agent to fully understand the boundaries of the property. If the agent doesn't know or if the party you are purchasing the farm from can't tell you, it may be worth the expense to have a surveyor establish all boundary lines around the farm before you agree to the purchase. All farms have deeds that contain legal land descriptions identifying the exact boundaries and a qualified surveyor can use these documents to determine all property lines.

If the lines seem to be well established, such as the line fences being intact, you may feel reasonably satisfied that the land you are purchasing is the one described on the deed. If a portion of the farm has been sold in recent years, you may want assurance that the new lines are identified in the deed and that they are the boundaries for the property you are buying.

Walking the fence rows before agreeing to a purchase will give you a valuable perspective of the farm beyond looking at plat or survey maps and photographs of the property. Walking the fences will also give you a sense of proportion and provide you an opportunity to inspect their condition and make note of problem areas that need attention.

Many old fences have been removed in recent decades to accommodate larger field equipment and to till the fence lines to keep weeds, small shrubs, and trees from growing. If fence lines are missing, it is important to know where they belong before constructing a fence of your own. Tearing down an improperly placed fence can be costly, time-consuming, and frustrating. Again, the expense to hire a surveyor to identify the exact property line before you build a perimeter fence may be a worthy expense.

Other aids that may assist you in understanding the dimensions of your property include aerial maps that have been developed by your county Farm Service Agency. Most, if not all, farming areas have aerial maps available, and they provide an overview of your farm not obtained from ground level. With their assistance, the relative boundaries of the fence lines can be seen and will give some indication of the lay of the land, which may be useful in handling any hills and valleys that the fence line traverses. You can request to see these maps prior to a farm purchase by visiting your county's FSA office and obtain copies once you have bought your farm.

One of the benefits of these maps is that they are developed in conjunction with farm programs and provide excellent guidance to determine which areas of the farm are best suited for different purposes. With the help of the FSA office, you will be able to identify those areas suited for hay

Some of the materials needed for constructing a good fence include T-posts, wire (woven and barbed), post hammer, shovel, a hand posthole digger, chain for pulling, claw hammer, and a manual wire-puller. Other items include pliers, wire cutter, tape measure, staples, smooth wire, wire clips, and a good pair of leather or heavy cloth gloves.

or permanent pasture, woodlands that may or may not be pastured, and areas suitable for cultivated crop production. These land assignments are made based on uses that will best protect the soil structure and return the greatest possible profit for each type of soil.

Planning Your Fence

If the perimeter fences are in good repair and appear to be free of holes or downed wires, you may need to do little construction work. However, if holes or breaks in the fence line exist, then it is wise to fix them to avoid escapes by your livestock or an easy entrance for predators.

If you need to construct a new fence, the key to building one successfully starts with a good plan on paper because alterations can be made quickly. Drafting a fence on paper also allows you to calculate the length of the fence and the amount of fencing materials needed before you drive in a post. It will also help you estimate the cost of your fencing project.

After you have developed your plan on paper, take a walk out to your fields and see if your ideas will work or if you need to make changes before you start. It is much easier to redraw your plan on paper than to tear down a partially constructed fence and start over.

Certain areas of your farm may be more suited for pastures than crops but are located a long distance from your buildings. In this case, you may need to construct one or more lanes, or pathways, for your livestock to gain access to those areas. A key rule is to build straight fence lines wherever possible. They are easier to construct, retain their tension for a longer period of time, and require fewer materials than curved fences. You may need to make curved fences in certain areas, but try to avoid these whenever possible.

Constructing your own fences will probably be less expensive for you than hiring out the work. Although you can hire fence contractors who will finish the work quickly (but at a greater cost), with a little experience and guidance, you can build a fence that will withstand the pressures placed on it by your livestock or weather conditions and last a long time.

Permanent and Temporary Fences

Good fencing protects and confines animals by providing a barrier to restrict their movements. A permanent fence that surrounds your farm is one of the best ways to protect your livestock investment. Besides establishing a fixed property line, perimeter fences are the last line of defense if your livestock happen to escape from designated pastures, feeding areas, or the small lots around your buildings. Perimeter fences will also keep your livestock from wandering onto highways and roads and protect them and the driving public from possible highway collisions. If your perimeter fences are intact it is unlikely your livestock will be able to invade your neighbor's property, thereby relieving you of possible financial liabilities due to destruction of property or crops. You cannot decide what species or stocking levels your neighbors may have, so maintaining good fences for yourself will give the same protection to your property by keeping their animals out.

Raising livestock will typically require two kinds of fences: permanent and temporary. Permanent boundary or division fences require different materials than fences for temporary lots. Permanent fences are intended to last many years with minimal repairs and should be constructed with sturdy and high-quality materials. Temporary fences are usually found inside permanent perimeter fences and are intended to subdivide fields into smaller areas called paddocks. They are built to last only a short time or to be easily moved from one area to another.

You should consider a permanent fence for those areas that will be used for pasture several years in succession. Farm ponds and other waterways should receive priority for permanent fences to control access or to allow access only for drinking. You may also consider permanent fencing for fields where cultivated crops are grown and your livestock are allowed access after harvest to graze the residue.

If it is not possible to construct a permanent fence around the entire perimeter of your farm, consider building those sections that will be most useful to your enterprise and plan to add the rest at a later date or by incremental construction of the rest.

Temporary fences are used for a few days, weeks, or months and can then be removed. Movable fences are less expensive to build than permanent fences and are quicker to set up and take down. However, they are less durable and may not be as effective, particularly if maintained by an electric current.

Easy mobility and light weight are two of the best advantages of temporary fencing. By being able to relocate them every year or several times

A review of the many types of fence available helps keep the costs and labor in line with your goals in erecting a fence.

within one year, temporary fences offer flexibility for pasturing livestock by expanding or contracting the size of the paddock or pasture. This flexibility allows you to accommodate any increase or decrease in the number of livestock placed within a pasture area, their relocation from pasture to pasture, and ease in developing different pasture rotation schemes.

Fencing Materials

Materials for permanent fencing that are most appropriate for livestock include woven wire, high-tensile wire, mesh wire, wood boards, or wire panels. Typically, barbed wire is not a good choice for fences for smaller livestock because it does not effectively deter predators and can cause injury to animals that happen to get caught in between several strands. However, barbed wire can be used at the top of a woven wire fence as a deterrent for other animals or predators; generally, one or two strands will be sufficient. Also, one wire can be laid at the bottom of the woven wire, along the ground, as an additional deterrent.

Temporary fences are typically made from materials that can be easily moved, such as a single metal wire, poly-wire, or wire tapes. Their mobility allows them to be rolled up on a spool and moved from one location to another with minimal effort then quickly unrolled to create a new temporary fence. Temporary fences are not considered for long-term use, although they can be if needed.

Your budget may have as much to do with the types of fencing materials you decide to use as anything else. Although differences in cost should be considered, the wiser consideration is purchasing materials that will last a long time. If your plans and goals change during the course of your farming career, you can use your well-constructed fences for a variety of livestock without making major alterations or additions. Though the cost of fencing materials is a major consideration, you should consider balancing that cost with the performance and time element involved. The initial cost of constructing a fence is spread over the life of the fence, and a well-constructed fence that requires minimal maintenance and repair over its lifetime is an excellent investment in the protection of your farm and livestock.

If you are planning to construct fencing around your entire property or fractions of it, you can consider using a combination of fencing types to suit different areas, possibly reducing the cost.

Woven-Wire Fences

Woven-wire fences consist of a number of horizontal lines of smooth wire held apart by vertical wires called stays. The height of most woven-wire fencing materials ranges from 26 to 48 inches. The spacing between

Woven wire is a good fencing material to use with goats or sheep. A woven-wire fence, when properly built, will maintain its strength and effectiveness for many years.

horizontal line wires may vary from 1½ inches at the bottom for small animals to 8 to 9 inches at the top for large animals. Wire spacing typically increases with fence height. Similarly, the spacing between the stays can vary, and you should consider using woven wire with stay spacing of no more than 6 inches for small animals. You can allow for 12 inches for large animals. The manufacturer's label attached to the wire roll will tell you the type and style of the roll by using standard design numbers and will help ensure you are buying the correct fencing product for your needs.

High-Tensile Fences

High-tensile electric fences consist of a single smooth wire that is held in tension between end-post assemblies. They are relatively easy to install, last for a long time, cost less than other types of fencing, and are easily adapted to specific needs. To be used with small animals, multiple strands—as many as five or six—are needed to keep them in and predators out because the bottom strands are more closely spaced than the top wires. It is important to keep the fence electrified to maintain the integrity of the wire. High-tensile wire can

No one kind of fence is best for all situations. Making a plan lets you select the best kind for each job, such as this alley fenced with high-tensile wire, which directs cattle to a pasture fenced with permanent barbed wire.

stretch if rubbed or pushed against by animals. This stretching will cause sagging of the wire and leave gaps for them to crawl through.

High-tensile wires are held on wooden fence posts with plastic insulators or can be threaded through plastic posts or metal posts with the use of insulators. High-tensile fences require strong corners and end braces to achieve adequate tension. One of the disadvantages of high-tensile fencing, particularly in climates that receive large amounts of snowfall during the year, is that the snow acts as a grounding agent and the electric current's effectiveness is diminished as it becomes buried under large snow drifts.

Electric fencing is used in areas separated from large portions of the field or pasture. The wire is attached to plastic insulators. This keeps them separated and ensures an unbroken electrical circuit.

Electric Fences

Electric fences offer enough flexibility for both permanent and temporary fencing needs. They are psychological barriers more than physical barriers, because the only thing that would keep your livestock within a designated area by using a single wire stretched across a field is the shock the animal receives when it touches the electric wire. The purpose of the shock is to scare or surprise the animal that touches the wire and alter its behavior to avoid the wire in the future. It is an effective and humane management tool.

Used as a movable or temporary fence, electric fences can be made with one, two, three, or more strands of smooth wire or a poly tape that has small wires woven into it. The poly tape is more flexible and easier to handle and move from one location to another than solid wire. Electrified metal wire, poly-tape, or high-tensile fences are typically energized by an electric controller that receives its source from a standard farm electrical outlet or a solar-powered pack. Both types need to be grounded to complete the circuit. Poor grounding is one of the leading causes of electric fence failures, and your electric fence must be properly grounded so that the pulse can complete its circuit and give the animal an effective shock. Always follow the manufacturer's instructions for grounding electric fences.

Electric fences can be useful in conjunction with permanent fences because they provide a second barrier of containment. This extra security is important if you have a permanent fence that may not be in the best condition but you do not desire to replace it. By positioning the wires near the permanent fence, you can accomplish the same result.

The electricity used in the wire is generally harmless to an animal or human, although it can cause momentary discomfort. A controller or charger regulates the flow of energy by supplying short pulses of electricity that travel along the wire. When your animal comes in contact with the wire, generally using their noses to investigate, they will complete the

Wire netting can be used in gateways or lanes. It has the advantage of providing an electric barrier and is easy to step over or open without touching the electrified threads.

An electric fence charger or controller is used to transfer and convert the flow of the electric current in your outlet to a series of short pulses that travel along the pasture wire. Controllers need to be correctly grounded to be effective in pasture use.

circuit from the wire through its body and through to the ground. This sudden shock or jolt will discourage further contact with the fence. Farm animals may require training when first using electric fences, and often they will not be aware of the fence until encountering a shock from the wire. It's important the fence charger be maintained for full-time operation in order for the temporary electric wire to be effective in keeping them inside the fenced area.

Solar power packs rely on the same principle as electric control devices but have the advantage of using an power source for which you do not pay. Solar packs derive their energy from the sun and deliver the current through the wire. Once fully charged, some packs have the capability of providing a low impedance current without the sun for up to two weeks. These may be useful in areas that are a distance from your buildings. Battery-powered charges are also an option, but they tend to have a shorter lifespan before they have to be recharged or replaced.

The effectiveness of an electric fence diminishes when the electric current no longer runs the length of the wire or when vegetation grounds the wire. In order to keep your electric fence in top condition, periodically check along the length of the fence to make sure the current is traveling through the entire fence. Also keep the grass underneath the fence trimmed so that no vegetation touches the fence to ground the current.

Wood Fences and Wire Panels

Using wooden boards for a fence is an option that provides a strong barrier and often an attractive feature to the farm. They are safe for animals and those working with them because of the lack of sharp points that may cause injury. One disadvantage of board fences is the higher cost to build and maintain them. It also requires more labor to plan and build a wooden fence than a wire one, although posts need to be set in either system.

Board fences need to be attached to wooden posts, but woven-wire, electric, high-tensile, and poly-wire fencing can be attached to steel or wooden posts depending on the purpose and preference of the farmer.

Board fences are usually made from 1- to 2-inch-thick, 4- to 6-inch-wide, 8-foot-long boards. A wood board fence typically will be 4½ to 5 feet high. By spacing the boards 4 to 6 inches apart, again depending on preference, you can calculate the amount of materials needed for your lot or pens. Because of the higher cost for board fencing, it is generally used in corrals or outside pen areas instead of the whole farm.

Wooden fences may be useful in small areas and around buildings, but they also have a shorter lifespan because of weathering effects and often begin to splinter, break, or rot after a number of years unless they are periodically repainted to repel the weathering effects.

Wire panels are an alternative to wood fences and, in some cases, to different electric fence types. Although similar in construction to woven wire, these panels are made of heavier metal material and are welded at the joints to provide a sturdy, long-lasting fencing option. Wire panels can also be used in close confinement areas, such as lots, small pastures, loading areas, and around buildings.

Wire panels are sometimes used as sturdy alleyways between buildings and can be easily erected and quickly taken down. The many and varied uses of wire panels provide the flexibility of a fence without it needing to be permanent or temporary.

Fencing Tools and Posts

If you are constructing a fence, the most important item you will need is a good pair of leather or heavy cloth gloves. Sturdy gloves are absolutely essential when working with wire as they reduce the risk of injury and severe cuts to your hands and fingers. Some of the tools needed to complete a fencing project include fencing pliers, a posthole digger, protective eyewear, a tape measure or reel, and a wire puller. When you have determined the type of wire that will work best for your situation and budget, you will also need staples, wooden posts or steel T-posts, and clips.

Fence posts are used to hold the wires apart or to keep woven wire fences erect and secure. They are commonly made of wood or steel; both have advantages and disadvantages. Wooden posts have the advantage over steel posts in strength and resistance to bending. Permanent fences often

The availability of cheap, plentiful local materials can make a big difference in the type of fence that's best for you. For example, an area with lots of industrial activity may offer surplus steel pipe and cable for fence building like this.

require decay-resistant fence posts. Wooden posts that have been pressure treated can last as long as steel posts. Wood posts come in varying sizes, so it is important that you use larger-sized posts for the corners and braces; smaller-sized posts can be used as line posts.

Wooden posts need to be long enough to support the fence height and the depth they are placed into the ground. A satisfactory post length is the height of the top wire above the ground, the depth of the post in the ground, plus 6 inches.

Steel posts have several advantages in that they cost less than a similar wooden post, can be driven into the ground more easily, and weigh less for easier handling. Steel posts generally are from 5 to 6 feet in length.

Fence Construction

The main ingredients to building a permanent fence with a long lifespan are solid end or corner posts, tight wire, and the use of good materials. Every fencing job has different requirements and each fence presents a slightly different approach. Like other construction and maintenance jobs around the farm, building a good fence requires proper techniques and a common-sense approach.

Clearing away brush, shrubs, and old wire will make the construction of a new perimeter fence much easier because it will leave you a clear path in which to work. If razing the area is not possible, you should at least try to clear the fence line of major obstacles that would impede your work.

If you are constructing a completely new fence, locating the corner or end posts should be your first step. (This step is not necessary if you are simply repairing a fence.) Locating the corner posts will allow you to plan your fence layout because you will be aiming for those corners as you unroll the wire. You may also want to determine placement of any gates if you decide to include them in your fence structure.

Steel T-posts are commonly used for fencing. They are driven into the ground with a post hammer. Plastic insulators can be attached to T-posts at varying intervals to hold horizontal wires. Use wire tighteners to keep them taut.

Gates and passageways should be located in the corners of fields nearest to the farm buildings. Placing them in corners makes it easier to move animals from one field to another and allows you to have your corner posts and gates in areas that do not break up the fence line.

Two of the most important contributing factors to a durable, long-lasting fence are well-set corner posts and tightness of wire; both are dependent upon each other. When wire is stretched, the pulling force on the corner post may reach 3,000 pounds. Winter cold can cause contraction of the wire, which can increase the pull to 4,500 pounds. The corner and end post assemblies must be strong enough to withstand these forces or the posts will slowly be pulled out of the ground enough so that the wire loses tension.

If you need to replace corner or end posts you should set them at a depth of 3½ to 4 feet. In colder climates, the shallower the depth, the more likely the post will eventually work its way out of the ground due to ground heaving in spring. A post moving vertically, even 2 inches, will cause the fence to slacken.

Once you have determined the approximate fence line, you can lay down a single strand of wire from one end post to the next. When tightened, this will provide a straight line for placing your wooden or steel t-posts at appropriate distances. Once the posts are set, you can unroll your woven wire and then proceed to stretch it with the use of a woven wire stretch assembly. This method will keep the wire from stretching out of shape and evenly stretch the entire roll. However, be aware that during this tightening the wire will have a tendency to turn over on itself. Care should be taken so that the wire remains flat to get an even stretch.

Depending on the length of your fence, the spacing between wooden line posts is normally 20 to 25 feet, and the spacing between steel posts is

Wood posts make great stabilizing anchors for fences. They are particularly useful when placed on hilly terrain or at corners. They can be used exclusively in a fence line or spaced between steel T-posts to help strengthen the fence.

Once the old wire has been completely removed from the fence line, it can be rolled up and sold for scrap metal or recycled. Old wire is typically not used again for making a new fence. Check each steel T-post you remove, as they may be good enough to use again.

normally 5 to 8 feet. If the total length of the fence is greater than 650 feet between corner posts, it is advisable to insert a braced line post every 600 to 650 feet. This brace will help keep tightness in the fence over long distances. Braced line posts are also very useful when the fence is made over rolling land or hills. Once the layout for the perimeter fence is finished, you can plan for any interior fences, corrals, waterway barriers, or other enclosures.

If you are creating lanes for your livestock, it is best to locate them in the driest areas possible, such as along a natural ridge or some other higher land feature. This positioning will allow you options for livestock usage if your future plans change.

Questions about Fencing

Many booklets that explain different fence construction options are available through county agricultural extension offices or fence manufacturing companies. Initial costs of fence construction are usually high, but programs through land conservation agencies that may help offset some of these costs when applied to certain management practices. Check with your county agricultural extension office to learn more about available programs.

Positioning all T-posts an equal distance apart between the wood line posts will help solidify the strength of your fence. With the end and brace posts set and the wood and T-posts in place, you are now ready to roll out your woven or barbed wire and attach it to the T-posts with wire clips. Your line fence is now properly built and will provide you with many years of service.

A well-maintained tractor will give you many years of good and trouble-free service. Regular tractor and equipment maintenance should be part of your farm schedule.

TRACTORS

Adapted with permission from *How to Keep Your Classic Tractor Alive* by Spencer Yost (Voyageur Press, 2009)

Maintenance is key to keeping your tractor running as it should. Proper and routine maintenance will minimize the need for repairs. The maintenance procedures are relatively simple, and the tools required to do a good job are not terribly expensive. Older tractors will likely require more attention than newer models. If you are mechanically minded, the time you spend in maintaining your tractor may be very rewarding. Even if you don't fit that category, maintaining a tractor can be a great learning experience.

Goals of Maintenance

The primary goals of any tractor or equipment maintenance are increased longevity, improved productivity, and continued assurance of the fitness of a machine for its purpose. Regular tractor maintenance increases longevity by preventing wear, which in turn keeps precision engine parts synchronized and within tolerances. Maintenance improves productivity by assuring the tractor's availability when needed and helping to reveal impending repair issues early while they are often easier and less costly to address.

Maintenance is the reason many old tractors are around today as well as other equipment. Setting a schedule, developing an approach, and making routine tractor maintenance part of your yearly schedule will make a major difference in its fitness, safety, and longevity. Issues such as storage when not in use, lubrication products, and other disposable supplies will be part of this review.

Some tractor maintenance procedures will be simple, such as greasing fittings. Others, such as timing the engine or adjusting the valvetrain, will be more involved. None are outside the scope or ability of anyone who wants to learn and try.

General Care and Scheduling

Tractors are quite rugged and certainly don't need to be pampered, but you need to routinely do certain basic procedures to ensure consistent performance and longevity. First, store the tractor indoors when it is not in use. Occasional out-of-doors storage won't hurt it, but leaving the tractor unprotected regularly will lead to weather damage to the sheet metal, tires, hoses, wiring, hydraulics system, and other components.

A general rule of thumb in tractor care is to inspect it thoroughly before each use. Inspect the tractor well enough to identify loose fittings, tire damage, cracked or leaking hoses, and other problems that can lead to trouble if they are not corrected.

To determine the best maintenance schedule, consult your tractor's owner's manual. A service or operator's manual will tell you how often to change the oil, transmission fluid, and hydraulic oil; when to lubricate certain components; and what style and type of fluids to use.

Lubrication

The majority of all maintenance procedures boils down to lubrication, cleaning up after it, or changing a fluid that lubricates. To apply grease to a lubrication point on a tractor, a one-way greasing valve called a "zerk" or grease fitting allows the grease to enter but not leave. Knowing where all these fitting are on your tractor is important, and applying grease may be the first maintenance procedure you will perform.

Most operator manuals will have drawings or photos outlining every grease fitting on your tractor and what type of grease to use if the fitting does not take ordinary grease. To apply grease, first clean the fitting completely. Then fit the tip of the hose of the grease gun over the grease fitting and pump the grease gun. The hose, if kept straight, will not come off while you pump grease, but it will be loose and easily removed when you are not.

Fasteners and Fastening

The proper way to tighten a bolt or screw is not always self-evident. Things like torque specifications, thread additives, and parent material composition need to be taken into account. When reassembling parts, you should try to clean out the threads of the fastener and parent material, and replace the fastener if it is heavily rusted.

Proper tightening may require a torque wrench. Bolts such as cylinder head bolts, bearing bolts, and similar bolts should never be tightened without a torque wrench. The torque wrench will tell you, in the units of foot-pounds, how much force you are applying to the fastener. Comparing this to the manufacturer's guidelines will help you to apply the proper amount of preload to the bolt.

Engine Maintenance

All engine components are high-precision assemblies subject to extremes of temperature and stress. Maintenance is critical to the longevity of your engine.

The most important thing you can do for your engine is to keep it properly filled with fresh, frequently changed oil. Your owner's manual will state what oil grade to use, how much will be required, and the type of filter to replace. It will also indicate the drain plug location. Always remember to replace the drain plug tightly once you have finished draining the old fluid to prevent new oil spillage and leakage.

Cooling System

Radiator fluid should be changed once a year on regularly used tractors. The fluid should be a 50/50 water/ethylene glycol mixture. This mixture will protect against freezing and boiling and will help to prevent further rusting and scaling of the cooling system. Check to see that the fluid level is within the range specified in the owner's manual. Refill it with mixture as needed.

TRACTOR QUICK TIPS

- The owner's manual should be your first source of information.
- Never store a tractor where it is exposed to the weather for an extended period of time.
- Always clean every grease fitting before use and wipe away excess grease.
- Regardless of type of fluid used, try to drain fluids when they are warm (but not hot). The old fluid will drain more completely.
- Repair or maintain primary, engine, and hydraulic systems in locations where dust and dirt are controlled.

SUGGESTED SCHEDULE OF MAINTENANCE

	HEAVY USE (More than 100 hours a year)	**LIGHT USE (Less than 100 hours a year)**
Oil change	Every 100 hours or at least every 6 months	Every 6 months
Chassis lubrication high-speed joints	Should be lubed every use, (spindles on mowing decks, etc.) otherwise once a week	Once a month
Air cleaner (wet oil bath type)	Once a week (daily in dusty or chaff-filled environments)	Every oil change
Air cleaner (dry paper type)	Once a month (once a week in particularly dusty or chaff-filled environments)	Once a year
Fuel filter or sediment bowl cleaning	Once a month (more frequently if gas tank is rusty)	Once a year
Fluid level check	Every use	Every use
Tire check	Every use	Every use
Tune-up	Each season	
Dressing	Once a year	
Spark plug gaps	Once a month	
Coolant change	Once a year	Every other year
Hydraulic fluid change	Every year	Every other year
Transmission fluid change	Every year	Every other year
Brake and clutch	Once a year or as needed	Once a year
Linkage adjustment	As needed	
Engine accessory drive belt	Dressed and tightened twice a season	Dressed and tightened once a year
Repack wheel bearings	Once a year and at every tire change and as needed	Every three years
Inspect structural bolts	Every month	Once a season
Bearing preload adjustment	Every season	As needed

Some general notes:

- *If an operator's manual can be found, follow your manufacturer's recommendations.*
- *These recommendations assume covered storage. (Lubrication in particular should be done more often if the tractor is left out in the weather.)*
- *This assumes the tractor is operated in normal agricultural conditions. Particularly dry and dusty conditions require more frequent maintenance.*
- *These are minimum maintenance intervals. Performing any of them more often is encouraged.*

Be sure to inspect for leaks from time to time as puddles on the floor or ground may indicate deficient hoses or simply fluid that has escaped through the overflow because of excessive boiling. Also be sure the overflow tube is unobstructed.

Part of regular maintenance on cooling systems includes cleaning chaff and trash from the cooling fans of the radiator and straightening bent fins.

Depending on the age of your tractor, the thermosiphon cooling system may or may not have a thermostat, which is a device that senses the temperature of the cooling system so that its temperature is maintained near the desired set point. The thermostat maintains that balance by switching heating or cooling devices on or off or regulating the flow of a heat transfer fluid as needed. To check the operation of the thermostat, simply check for the presence of heat either at the top of the radiator or at the hose running to the top of the radiator. Heat can be present after the tractor has worked for a while. The lack of heat means the thermostat isn't functioning properly or was removed by a previous owner and needs to be replaced. Also check the radiator cap to make sure it provides an adequate seal when fully tightened. Replace any worn or defective radiator cap as these need to withstand differing amounts of pressure during tractor use.

Radiator

The radiator helps cool the engine by allowing air to be drawn through it by the radiator fan. The cool water circulates around the engine block

BOLTS, NUTS, AND OTHER FASTENERS

Grade 8 fasteners: Cylinder heads, structural bolts (bolts that hold main structural pieces of the tractor together), any bolt or nut with a high torque specification (greater than 125 foot-pounds), axle nuts, and rim-to-wheel bolts. You can almost never go wrong using a hard bolt in place of a soft bolt. Use this grade if unsure.

Grade 5 fasteners: The most common grade of fastener used. About 90 percent of all bolts on your tractor can and should be this grade.

Less than Grade 5: For light-duty and decoration purposes. Do not use this fastener in any situation where the bolt has to be relied upon to prevent injury or property damage.

Shear bolts: These bolts are designed to be very soft and break or shear when a heavy load is applied. This is done for safety reasons so equipment is not damaged nor people injured when shock load occurs. **Never** substitute regular bolts for shear bolts.

drawing the heat back to the radiator where it is cooled. Leaks can occur in radiators as the warm fluid increases in pressure. Always check the fluid level in your tractor's radiator prior to starting it. A consistent need to refill indicates a slow leak that must be addressed. Running low on radiator fluid will cause the engine temperature to rise because it can't be sufficiently cooled. Maintain proper radiator fluid levels with a mixture that keeps the water from freezing during cold temperatures. Air flow is curtailed if the radiator water is frozen.

Radiators may be removed to be cleaned if necessary. Be sure to drain all fluids before removing or working on it. Many antique tractor radiators are three-piece systems bolted together, and not a solid assembly. Leaks often occur at these seams, so if your radiator is leaking, check the seams before you get a costly radiator recore done.

The radiator typically has an external screen or grill that must be kept free of debris, such as plant material, insects, or pollen, to allow sufficient air flow. Remove anything that blocks the air from entering the radiator.

Fuel and Air Delivery

The carburetor and fuel delivery system must be maintained. Fortunately, very little maintenance is needed. On gas engines, the maintenance includes replacing the fuel filter, adjusting idle speed, adjusting the fuel air mixture, and cleaning or replacing the air filter.

Adjusting idle speed requires a tachometer. Install the tachometer, idle your engine, and then turn the idle-speed screw until the tachometer reads the required revolutions per minute (rpm). Once that is done, you can adjust the idle-mixture screw. This control adjusts how much fuel enters the air stream going into the engine. This particularly affects low-speed engine performance. Some carburetors do have high-speed mixture-adjusting screws. Their adjustment is simple and requires an owner's manual to find the proper initial setting and the proper adjusting procedure.

Diesel fuel systems are difficult for the average owner to service alone outside of a couple of small maintenance tasks. First, regular attention should be paid to the system's filters. At least one, usually two, filters separate and remove water and contaminants from the fuel. Since the diesel fuel pumps, metering apparatus, and injectors are high-precision components, make very sure that your fuel, fuel tanks, and filters are kept very clean.

Exhaust System

The exhaust system on most tractors is simple and requires very little

The most common exhaust design on older tractors is the straight-up exhaust. Your exhaust muffler should be long enough to put the outlet well above the tractor operator, so the operator doesn't breathe the exhaust. All straight-up mufflers should be topped with a rain cap.

maintenance, usually only an inspection. You are looking for exhaust leaks for the most part. Exhaust leaks can emit sparks, which can be very dangerous and must be prevented. This is especially true if you use the tractor around buildings, fuel, hay, or other combustible materials. The common places exhaust leaks are found is at intersections of manifolds and exhaust pipe, the intersection of the muffler and exhaust pipe, and from around the exhaust manifold gasket. Occasionally, the heat, age, and rust eventually erode the exhaust manifold to the point where holes develop in the manifold itself. Look primarily at the corners and underside of the manifold to find these trouble spots.

Electrical

The lighting, starter, battery, and generator are all part of the tractor's primary electrical system. The portions of the electrical system that generate higher voltages for the formation of a spark at the spark plug are called the secondary system. Examples of secondary components include the coil, the high tension side of distributors, and spark plugs. A diesel engine does not have a secondary system.

The primary electrical system requires little maintenance, and most of that involves caring for the tractor's battery. Most batteries sold today are enclosed although some batteries can be opened to inspect the cells. These types of batteries will need to be inspected and must be topped off with distilled water on a regular basis and should be kept charged up during periods of nonuse.

The battery connections should be kept clean and any scaling or corrosion should be neutralized with baking soda when needed. Be sure not to get any of the baking soda solution in the cells of the battery. The other areas of maintenance for the primary electrical system are keeping connections clean and tight and inspecting the condition of the wires from time to time. Also, you should perform a voltage check of the charging system.

The secondary electrical system is that part responsible for creating a spark at the spark plug. This part of the electrical system, also called the ignition system, requires much more maintenance than the primary system. The purpose of the ignition system is to create a high-voltage current and

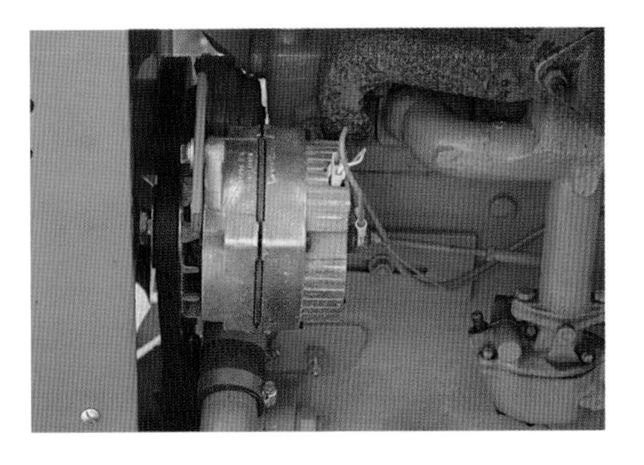

Alternators have one advantage to generators in that they charge at a constant rate, even at low engine rpms. Most alternators operate on a 12-volt battery system, whereas many old tractors have a 6-volt generator, so you'll have to get a new battery if you want to make this upgrade.

send it to the spark plug at the right time. Maintaining this system is a bit more difficult, but these procedures are more important than most.

Magnetos in older model tractors often do not fail completely, but will fail intermittently, making them difficult to troubleshoot. However, the procedures for maintaining them are easy. They include gapping the points, maintaining spark plugs and coil wires, and timing the engine. First, clean the magneto away from the tractor. Then lubricate it and replace it.

Tuning up your tractor involves first inspecting, cleaning, and gapping the spark plugs. To clean the plug, use a brake- and carburetor-cleaning solution. Then check the ignition point gap. The points are found inside the distributor underneath the distributor cap. The points open and close as necessary, creating and breaking the magnetic field in the battery coil that ultimately creates the spark. How long these points stay closed is important, and setting the gap ensures that the current that is building in the electromagnetic field stays on for the proper length of time.

Timing the engine is the next step in tuning up the ignition system. Timing involves the spark at the spark plug occuring at the proper instant during the engine cycle. The spark must be initiated slightly before the piston is at the top of its compression stroke. All engine timing is expressed in the number of degrees before top dead center (BTDC). This range is usually about 2 to 8 degrees, but it can be slightly more than that. Your operator's manual will have the proper timing specifications and procedures.

Fluids

Several systems in your tractor use fluids to operate correctly, and these systems should be checked periodically. These fluids include the engine oil, the transmission fluid, radiator coolant, and hydraulic oil.

Tractor usage is measured in hours, not miles like automobiles. Leakage can cause failure of any of these systems, so it is important to check the hoses through which these fluids flow for any cracks or defects and replace those that show signs of wear or weakness.

Each fluid system will have a dipstick, gauge, or reservoir that measures the level of fullness. Check your operator's manual for the locations of these gauges and specifications for each system. Then fill each to the appropriate level with the recommended type of fluid. Some fluids are different for the engine, transmission, and hydraulic systems, so be sure to use the correct one.

It is important to maintain correct fluid levels in all tractor systems. Failing to do so may cause the system to fail and require expensive repairs such as low engine oil levels, which may cause excessive wear on the internal gearing. In some cases, a system failure may create a hazardous or life-threatening situation, such as the loader on a front-end loader falling due to inadequate pressure in the hydraulic system hoses.

Lubrication Points

Bearings and bushings throughout the tractor require periodic maintenance that typically means cleaning bearings and lubricating them. Some bearings may come from the factory permanently lubricated and therefore do not need lubrication. Bushings typically receive their grease through a zerk fitting. Your tractor's manual will identify every lubrication point that needs your attention.

Chains, either drive chains or power transmission chains used to couple shafts, usually require light oiling, but some are greased. Whether you should oil or grease them will be outlined in the owner's manual.

Metal-to-metal contact areas need lubrication. These components rub or mate together, yet do not use bushings or bearings to control wear or positioning. Occasionally they have zerk fittings for the application of grease, but often a light coat of oil occasionally applied is all they need. Typical examples would be the pivot points for levers, throttle linkage joints, and steering column braces.

Wheels and Tires

Rims, wheels, and tires require periodic maintenance. Rear tractor tires are one of the most expensive parts of a tractor so proper maintenance makes good sense. The tires should be cleaned thoroughly once a year, and the tire pressure should be checked frequently. Rear tires normally have between 12 and 20 pounds per square inch (psi), while the front tires may have up to 32 psi.

Many rear tractor tires are filled with ballast, which is a liquid to help pull an implement where maximum traction is needed. The liquid is usually water-based with an antifreeze solution such as ethylene glycol added. If this solution leaks out of the tire, replace the solution and clean the rims and wheels after refilling.

Rear wheel width adjustments can be made to the adjustable rim, which is often known as the spin-out rim. By loosening locking bolts, you can rotate the wheel in or out to get the desired width. Be sure to use sturdy blocking if lifting the wheel off the ground for adjustment.

To check the tire pressure, use a tire pressure gauge placed over the valve stem and hold it there firmly until the pressure is registered.

Periodically check the lug nuts on the rear wheels. They can work loose if they have not been torqued properly. You may need to reset the wheels for field work if it requires different wheel width settings. Some

Tractor tires have evolved into various tread designs over the years. (The tread shown is on an old tractor—this design is hard to find these days.) Good tread equals good traction and safe use. Replace any tires where the tread has worn down so that they slip when first starting to pull their load.

field equipment may work better with a narrow wheel width; others need to be set at the widest width. If you need to move the wheels be sure the proper torque has been applied when tightening the lug nuts again.

Belts and Hoses

Most modern tractors have very few belts. The most common belt is the fan belt, which helps run the cooling system. Check this belt for cracks or other weaknesses that may cause failure. If the fan belt breaks while you are operating your tractor, turn off the tractor immediately. Running an engine that cannot cool itself will create a high internal heat level that may crack the engine block. It may be inconvenient to stop somewhere in the field away from the farm, but it is less expensive than having to replace a major component of the tractor.

Check all belts on your engine and replace any that are frayed, cracked, brittle, or loose. Pictured here is a belt-driven aftermarket system fitted onto an old tractor.

High pressure hoses such as those involved with the hydraulic system need to be checked periodically for stress, cracks, leaks, and normal wear. Loss of pressure can result in steering problems, rear lift-arm failure, and a dangerous situation if a loader is raised when a failure happens.

Besides the hoses, check the fittings or connections at the ends of the hoses to be sure they are not leaking. Tighten or replace any fittings or connections that leak or check the seals to make sure they are working properly. The types of belts or hoses used will be listed in the owner's manual.

Brakes

Modern tractors have mechanical brakes. These types of brakes are operated by a linkage and cam system instead of a master fluid system. They are located on the rear axles and work independently, which allows them to be used to steer the tractor in turns or to reverse the direction of travel.

The brake fluids may or may not be part of the hydraulic system. Check your tractor's manual to identify where the fill points and fluid level gauges are located.

It is important to maintain proper brake fluid levels because brake failures can have catastrophic effects.

Gauges

Three major gauges need to be checked often: water temperature, oil pressure, and electrical. Temperature gauges are marked with a normal operating range and an indication area if the engine is exceeding that range. Most tractors have an internal temperature range of between 180 and 220°F. If the internal temperature exceeds that range, the engine can overheat and cause severe damage to the pistons, piston rings, engine block, and other components.

The oil gauge measures internal oil pressure and keeps the mechanical parts from rubbing on other parts. High oil pressure will put an oil film between these parts. A loss of oil pressure can cause a lack of lubrication from the oil film. This imbalance can cause the parts to heat very quickly and the engine to lock up. Make sure the oil gauge is working correctly and doesn't contain moisture drops, cracks, or dirt that can indicate a malfunctioning gauge.

The electrical system is monitored by the ampere gauge that shows the voltage reserve of the battery. A positive charging rate indicates the battery is being charged as it is being used. A negative rating means the battery is not being charged and could indicate a weak battery, improper or faulty wire connections, or an electrical system that is drawing too much from the battery. Watch for any negative rating as this indicates a problem that you will need to address.

Common maintenance tasks on tractors are oil changing and filter replacement. All engine components require correct levels of oil to work properly. As engines work, they create minute bits of debris or contaminants that are collected in the filter. Regular replacement of oil and filters will help keep your engine running efficiently. If you have an antique tractor like the one pictured, you'll also need to take off the band around the starter brushes and clean the brushes of dirt and dust.

Filters

Several filters in any tractor need attention: including the air, oil, and fuel filters. Check the air filters often and replace those that have significant dust deposits. Tractors often work in areas that are very dusty. If the field conditions warrant it, you may have to check and clean the air filter daily or weekly. Never wash an air filter as the moisture will block air flow. Replace any air filter that cannot be cleaned properly or has been damaged.

In old tractors, there's usually a glass bowl found under the fuel tank. Inspect this bowl every season. Dirt or sediment leaves the fuel tank and is trapped in the glass bowl to prevent it from reaching the carburetor. Turn off the fuel flow with the small valve seen at the top left of the sediment trap, loosen the nut at the bottom, and swing the wire sling out of the way. Remove any dirt or sediment found in the bowl before refitting it into the holder.

Oil filters should be replaced every time you change the tractor's engine, transmission, or hydraulic oil. Replace old filters with ones recommended in your tractor's manual.

Fuel filters keep dirt and debris that may have entered the fuel tank from advancing to the engine. Older tractors may have sediment bowls along with fuel filters. Each time you replace a fuel filter you should also clean out the sediment bowl if your tractor is equipped with one. Be sure to turn off the fuel line before loosening the tightening nut on the fuel bowl.

Records

Keep good records of all the servicing you do to your tractor. This includes dates, tractor hours, filters, lubrications, replacement parts, and anything else by which you can measure its performance. Regular service records will also provide an accurate maintenance schedule should you decide to sell your tractor.

As the owner of a small farm, you may want to teach yourself to weld in order to repair your own machinery, equipment, and tools. This chapter is an introduction to welding—the tools, the gear, the safety issues, and the basic processes. Before you invest in tools or gear, though, take some welding classes through your local technical college or elsewhere: welding is an expensive undertaking and a hazardous one.

WELDING

Reprinted from *Welding Complete* (CPI, 2009)

Welding is all about heat. When you weld, you use heat to melt separate pieces of metal so they will fuse to form a single piece. Your ability to control the heat generated by the flame or arc determines the quality of your welds and cuts.

Certain terms are used to describe the heat and action of all the welding processes. The parts being welded together are referred to as the *base metal*. Additional metal, called *filler*, is often added to the weld. The *molten puddle* is the area of melted base metal and filler metal that you maintain as you create your weld. To have fusion of metals, the base metal and filler metals must have the same composition. Methods for joining metal without fusion are called *soldering*, *brazing*, and *braze welding*. These methods can be used to join similar or dissimilar metals.

Oxyacetylene welding and cutting use flames to generate the heat to melt the metal. Shielded metal arc, gas metal arc, gas tungsten arc welding, and plasma cutting use an electric arc for heat generation. With oxyacetylene welding, you have time to watch the puddle develop as the metal turns red, then glossy and wet looking, then finally melts. With the arc processes, the puddle forms quickly and may be difficult to see because of the intensity of the arc. It is important to have ample lighting and clear vision so you can watch the puddle and move it steadily.

Penetration of the weld is also a critical heat-dependent factor. A strong weld penetrates all the way through the base metal. Matching filler material size and heat input to the thickness of the base metal is important. It is easy to get a good-looking weld that has not penetrated the base metal at all and merely sits on the surface. At the opposite end of the spectrum from a "cold," nonpenetrating weld is burn through—where the base metal has gotten too hot and is entirely burned away, leaving a hole in the base metal and the weld.

Heat distortion is a byproduct of all welding and cutting processes. It is obvious that applying a flame to metal in the oxyacetylene process makes the metal hot—and the electrical arcs are actually four to five thousand degrees hotter than the oxyacetylene flame. When metal is heated, it expands; when it cools, it contracts. If not taken into consideration, this expansion and contraction may cause parts to move out of alignment. For this reason, tack welding and clamping project pieces are critical to successful welding. It is also important to match the welding process to the base metal thickness. For example, shielded metal arc welding (SMAW) is generally not used on materials less than ⅛ inch thick. The process is too hot and too difficult to control on thin material. On the other hand, gas metal arc welding works well on very thin sheet metal, if you are able to adjust the machine to a low enough voltage.

Protecting the molten puddle from oxygen is also an important part of welding. Oxygen makes steel rust and causes corrosion in other metals. The exclusion of oxygen from the weld when it is in the molten state makes a stronger weld and is accomplished in a variety of ways. A properly adjusted oxyacetylene flame burns off the ambient oxygen in a small zone around the weld puddle. Gas metal arc and gas tungsten arc welding bathe the weld in an inert gas from a pressurized cylinder. Inert gas keeps oxygen away from the molten puddle. Flux cored arc welding and SMAW use fluxes in or on the welding filler metal. When these fluxes burn, they produce shielding gases and slag, both of which protect the weld area until it has cooled.

Welding well is difficult and takes years to master. It's a good idea to talk with a welder and have him or her evaluate some of your welds. Remember the safety of others is involved when you're making repairs to tools and farm equipment, so take the time to ensure that any project you make is safe.

Welding safety equipment includes: (A) safety glasses, (B) particle mask, (C) low-profile respirator, (D) leather slip-on boots, (E) fire-retardant jacket, (F) fire-retardant jacket with leather sleeves, (G) welding cap, (H) leather jacket, (I) leather gloves with gauntlets, (J) heavy-duty welding gloves, (K) welding helmet with auto-darkening lens, (L) welding helmet with flip-up lens, (M) full-face #5 filter, (N) full-face clear protective shield.

This section is intended only as an introduction to some of the tools and processes of welding. If you wish to further your welding skills, many community colleges, technical schools, and art centers offer welding classes. Such classes are an ideal way to learn the basics of welding, proper techniques for each process, and welding safety.

Safety

When safety measures are overlooked or ignored, welders can encounter such dangers as electric shock, overexposure to fumes and gases, arc radiation, and fire and explosion, which may result in serious or even fatal injuries. Preparations for welding—grinding, burnishing, and sawing—are also potentially dangerous. Familiarize yourself with the American National Standard Safety in Welding and Cutting practices (see Resources, page 308). Always read, understand, and follow manufacturers' material safety data sheets (MSDS), which include specific instructions and recommendations for equipment and product use. Also follow general welding safety rules.

Fumes. Welding produces potentially hazardous fumes and gases. Always keep your face out of the weld plume and avoid breathing concentrations of those fumes. Welding indoors requires special precautions when considering ventilation and may require a ventilation fan, exhaust hood, or fume extractor. If you cannot provide adequate ventilation, an Occupational Safety and Health Administration (OSHA)–approved respirator or particle mask may be needed. If you may be pregnant or plan to become pregnant, consult your physician.

Burns. Hot sparks and flying slag can cause burns. Protect your hands, head, and body with natural fibers such as leather, wool, or cotton. Synthetic fibers like nylon and polyester melt when ignited, which causes serious burns. Wear leather slip-on boots and pants that fall over the top of the boots. Cuffed pants, frayed edges, and pants with holes are fire hazards. Wear a welding cap and keep hair tucked away.

Arc burn. Welding arcs produce ultraviolet and infrared light. Both of these can damage your eyes permanently, burn your skin, and potentially lead to skin cancer. A welding helmet with a filter lens protects your eyes and face; long-sleeve shirts and long pants protect your skin. Remember to use welding screens and have an extra helmet on hand for observers.

Fire. Remove flammable items such as lumber, rags, drop cloths, cigarette lighters, and matches from the work area. Do not grind or weld in a sawdust-filled shop because the sparks can ignite airborne dust and fumes or ignite flammable materials. Mount an ABC-rated multipurpose dry chemical fire extinguisher and first aid kit in the work area. Periodically check your welding area up to one half hour after welding to make sure no sparks have found a place to smolder.

Explosion. Nonflammable gases such as carbon dioxide are stored in high-pressure cylinders that can be dangerous. If regulators are not installed, keep the cylinder's protective cap on and keep cylinders chained or strapped. Never use cylinders as rollers or supports, and never weld protective caps to anything. Shut acetylene and oxygen cylinder valves if you're away for more than ten minutes. Cylinders must be transported right side up and chained, even if empty. Similarly, never weld or cut on any closed container, tank, or cylinder. Bring tanks to a welding specialist for evaluation and repair. Another explosion hazard is concrete. Tack welding is acceptable, but using a concrete surface for welds is dangerous due to the amount of water contained in concrete. Use fire bricks instead.

Electric shock. Be sure to insulate yourself from the work piece and ground using dry insulation. Always wear dry, hole-free gloves. Do not touch electrically "hot" parts or electrodes with bare skin or wet clothing. Only use electrode holders and cable insulation in good condition.

Other hazards include noise from grinding, sawing, sanding, and plasma cutting; laceration from sharp metal edges; electrocution during arc welding; and asphyxiation from inert shielding gases. Carefully read all manufacturers' instructions before starting any welding process.

Setting Up Shop

If you plan to weld on a regular basis, it makes sense to set up a welding shop. The primary concern with welding is containing the hot sparks or slag and the flammable elements while exhausting dangerous fumes. It is possible to weld outside, but not all processes allow that. Gas metal and gas tungsten arc welding require that the surrounding air be still so the shielding gas is not disturbed. All the arc processes must be done in dry conditions

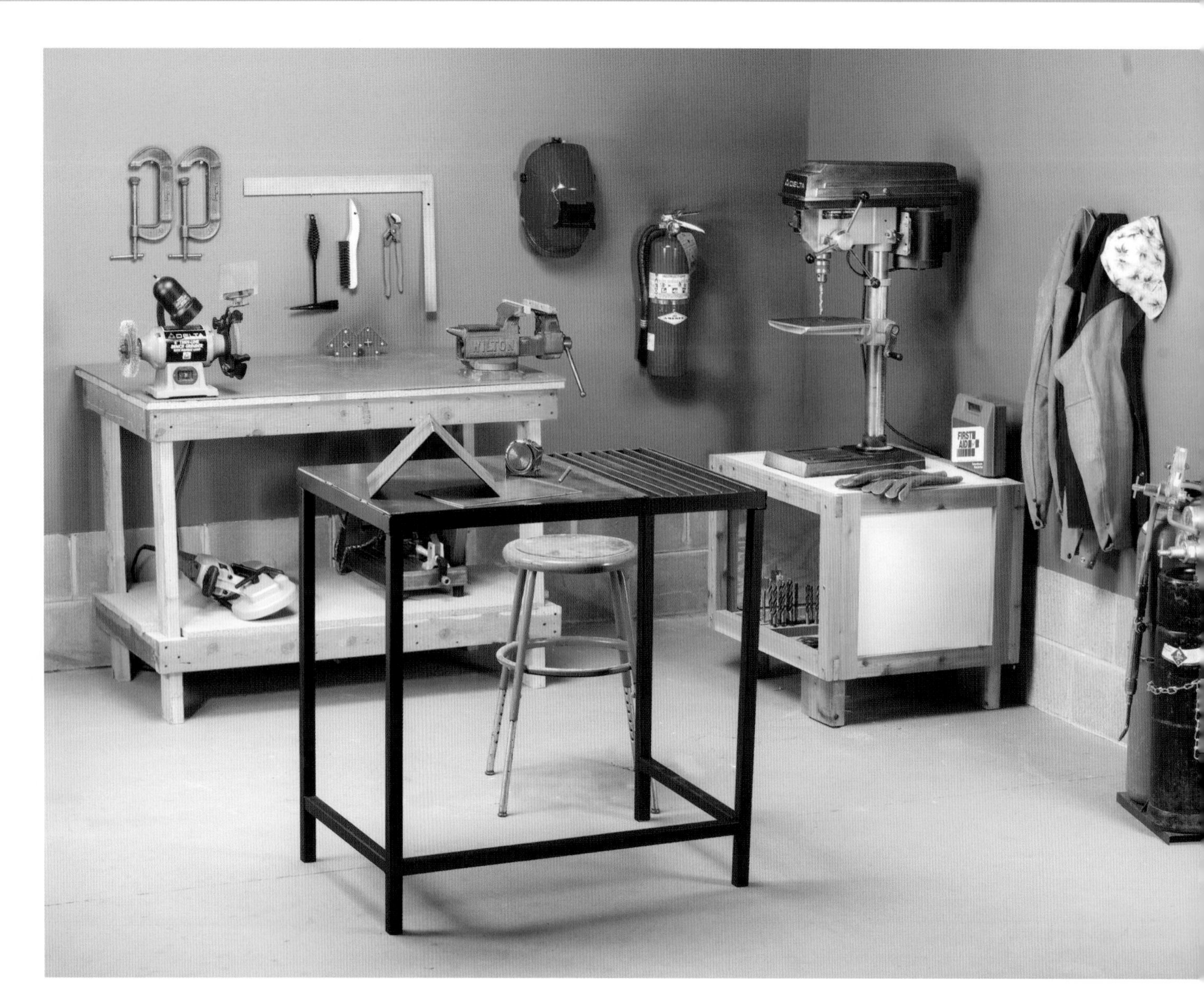

Shop space with a concrete floor and cement block walls is ideal for welding. Good ventilation is critical.

to prevent electrical shock. Cold metals do not respond as well as metals at 70°F and may not be weldable using certain processes. A heated garage or outbuilding is best suited for a welding shop. A basement is not suitable due to the dangers of fire and compressed gases next to living spaces. Additionally, your homeowner's insurance may not cover a welding shop if it is inside your living area instead of in a detached garage or shop building.

Tiny sparks and pieces of hot slag may scatter up to 30 or 40 feet from the source. If they come to rest on flammable materials, they may smolder, and given the right conditions, they can ignite the material. Always check your welding area one half hour after you have completed welding to make certain no sparks are smoldering. A wooden table covered with metal is not a good work surface, as the transferred heat may cause the wood to smolder. Any wooden jigs or clamping devices should be doused with water to extinguish smoldering embers or stored outside after use.

Power tools such as the reciprocating saw, angle grinder, portable band saw, and chop saw are useful when cutting and fitting metal parts to be welded. A drill press and metal cutting band saw are also useful tools for welding. Specific metal-working tools, including metal brakes and metal benders, are available for all sizes of metal. These can range from inexpensive sheet metal tools to expensive hydraulic tools.

Metal Basics

Different metals have different characteristics that affect their ability to be welded or cut with any of the processes described in this book. In general, only metals of the same type are welded together, because welding involves melting the base metal parts and adding melted filler metal. In order to accomplish this seam, the parts and filler must have the same melting temperature and characteristics. Dissimilar metals can be joined by brazing and braze welding because these processes do not actually melt the base metal.

Metals are divided into the categories of ferrous and nonferrous. Ferrous metals contain iron and will generally be attracted to a magnet. Cast iron, forged steel, mild steel, and stainless steel all contain iron, along with varying amounts of carbon and other alloying elements.

Mild steel, which has a low carbon content, is the most commonly used type of steel and the easiest with which to work. It makes up many of the metal items you commonly use, make, or repair, such as automobile bodies, bicycles, railings, and shelving. Adding more carbon to steel makes it harder, more brittle, and more difficult to cut and weld. High-carbon steel is called tool steel and is used to make drill bits and other metal cutting tools.

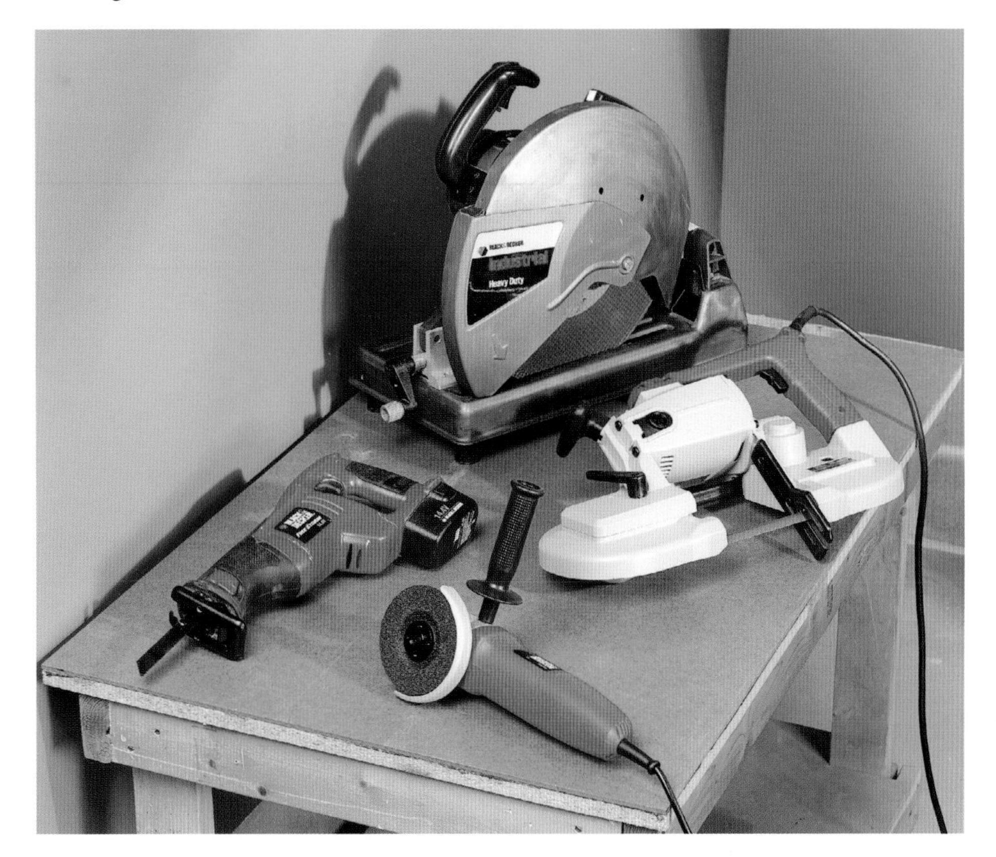

Power tools for a welding shop include: (A) reciprocating saw, (B) cut-off saw, (C) portable band saw, (D) angle grinder.

Other elements are added to steel to make a variety of alloys, such as chrome moly steel and stainless steel. These additions may make the steel nonmagnetic. Because stainless steel does not oxidize (rust) easily, it is difficult to cut with oxyfuel. When welding steel alloys, the filler material must be matched to the alloy to get a high-quality weld.

In addition to alloying, many metals are heat treated to improve characteristics such as hardness. Because welding and cutting would reheat these metals and destroy those characteristics, heat treating is an important factor to consider. Many vehicle chassis parts, including motorcycles and bicycles, are made of heat-treated metals.

Aluminum is a widely used nonferrous metal because of its light weight and corrosion resistance. Like steel, it is available in many alloys and is often heat treated. Aluminum is used for engine parts, boats, bicycles, furniture, and kitchenware. Various characteristics make aluminum difficult to weld successfully: it does not change color when it melts, it conducts heat rapidly, and it immediately develops an oxidized layer.

Repairing Metal

Performing welded repairs can be very tricky. It is important to assess your welding skills, the difficulty of the repair, and the intended use of the repaired item. Any structural or vehicle repairs, such as stairways, ladders, trailers, or truck chassis, need to meet the same safety standards as they did in their original condition.

The first step in considering a repair is determining why the item broke. If a weld was poorly executed, the repair might simply be to prepare the area

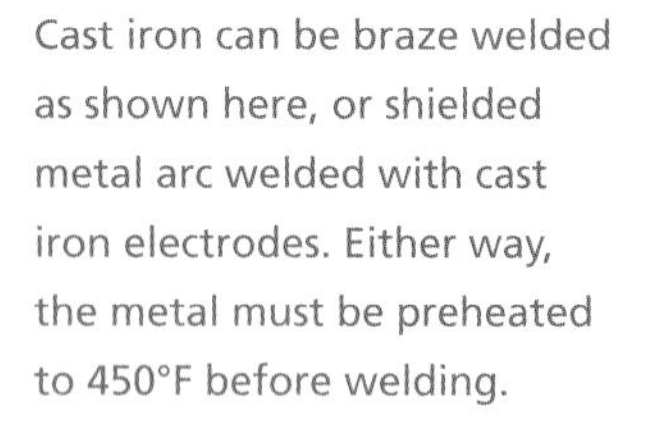

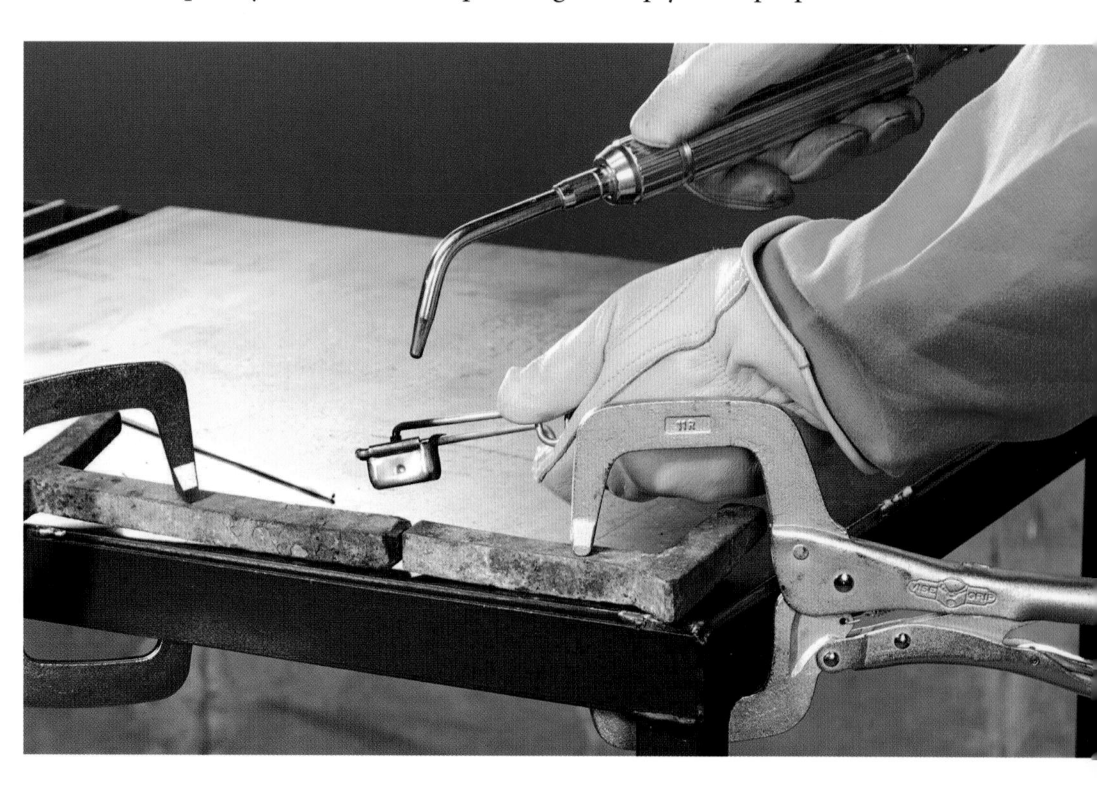

Cast iron can be braze welded as shown here, or shielded metal arc welded with cast iron electrodes. Either way, the metal must be preheated to 450°F before welding.

and perform a good weld. If a piece has broken due to metal fatigue, simply patching over the crack will only cause more cracks outside the patched area. Cast iron and cast aluminum may have broken due to imperfections or inclusions in the casting, or they may have gone through rapid heating and cooling that caused cracking.

The next step in a repair is determining the base metal. A magnet will be attracted to any metal with a certain concentration of iron in it, but stainless steel (which is sometimes nonmagnetic), mild steel, and cast iron require very different welding techniques. Aluminum is nonmagnetic and is discernible from stainless steel by its light weight—but which alloy is it? In the case of aluminum, some alloys cannot be welded. Unfortunately, many manufactured metal items are made of alloys and may have been heat treated. Without access to the manufacturer's specifications, it might be impossible to determine what the base metal composition is. It is important to understand the effects of welding on these materials before attempting a welded repair.

Once you have determined the feasibility of a repair on a particular piece, the piece must be prepared carefully. All rust, paint, and finishes must be removed from the area of the weld and, if you are arc welding, from an area for the work clamp. Grease and oil must be cleaned off. If the break is on a weld joint, the old weld bead needs to be ground down.

Mild steel is the easiest material to repair. Simply prepare the material as for any other weld, and weld using any of the welding processes.

Cast iron and cast aluminum need to be preheated before welding to prevent cracking due to temperature fluctuations. If the piece is small enough, you can put it in the oven and heat it to 450°F. Otherwise, use an oxyacetylene or propane rosebud tip. Temperature crayons, which melt at specific temperatures, are available for marking metal to be preheated. Cast iron can be brazed if the fit between the broken parts is good, or it can be braze welded or welded with shielded metal arc. Specific electrodes for cast iron are available. Cast aluminum can be brazed, braze welded, or welded with gas tungsten or gas metal arc welding.

Some aluminum is not weldable. If the piece to be repaired has been welded, you can perform a welded repair. If you can determine the composition of aluminum parts, match the electrode or filler rod to that. Some filler rods and electrodes are multipurpose and can be used on more than one alloy. Gas tungsten arc welding is the best choice for aluminum, but it can also be welded with gas metal arc, or it can be brazed.

Stainless steel comes in numerous alloys, and it is important to match the filler material to the base metal. Do not clean stainless steel with a steel wire brush, as the mild steel wires may contaminate the stainless steel. Stainless steel is best repaired with gas tungsten arc welding, although it can be brazed as well.

Cleaning Metal

A successful weld begins with a well-prepared piece of metal. Steel is not available in a pre-cleaned state, so you will need to clean all project parts by removing oil, mill scale, dirt, and rust. Although this can be a tedious step, a well-cleaned project lasts many years.

For any project that will be painted, it is important to thoroughly clean the entire part. For objects that will be allowed to rust gracefully outdoors, thoroughly clean the joint areas and allow the weather to clean the remainder.

The first step in cleaning is to wipe down the part with denatured alcohol, acetone, or a commercial degreaser to remove oil and dirt. The alcohol works best as it has little odor, unlike the degreaser, and it won't dissolve plastics, like the acetone will. Both acetone and degreasers tend to leave a residue that may diminish the quality of the final finish.

Once the grease, oil, and dirt have been removed from the surface, the mill scale (often found on mild steel) must be removed. Mill scale is a dark gray flaky oxide layer that forms on the steel as it cools. Cold-rolled steel often has a thinner layer of scale than hot-rolled steel, but even the thin layer needs to be removed before welding and finishing. Manufacturers use a pickling bath to remove scale; except for very small projects, it is unlikely

Apply denatured alcohol with a clean rag to clean dirt and oil from project parts. Protect your hands with rubber gloves.

that a home do-it-yourselfer would want to work with the hot acid bath needed for pickling. The remaining options are wire brushing, grinding, sanding, or sandblasting. A bench-mounted power wire brush is perfect for cleaning the ends of parts.

Unfortunately, it is not suitable for cleaning the remaining surfaces. For this you will need an angle grinder outfitted with a wire brush cup or flap sander. A drill or rotary tool with a wire brush wheel, wire brush cup, or flap sander also works well. When working with power wire brushes, it is crucial you wear appropriate protective gear. A full face mask and long sleeves protect your eyes and arms from wire fragments that are thrown from the brush. Heavy leather gloves will protect your fingertips from getting skinned. These tools operate at very high speeds and can quickly do serious damage to eyes and skin.

Basic welding skills are techniques that you will use again and again regardless of the type of welding you pursue. Described here are a few basic techniques, including mechanical metal cutting, shaping, brazing, braze welding, and SMAW.

Clean the mill scale off the ends of mild steel parts with a bench-mounted wire brush. Wear heavy leather gloves to protect your fingertips.

Mechanical Cutting

For some projects, cutting the metal is the most difficult step. A variety of tools exist for cutting metal, but the thicker and larger the stock, the fewer your choices. If you want to build muscle, a hacksaw is great for tubes, rods, and bars, but a large project will soon have you wishing for some more powerful cutters. Cutting thin-gauge sheet metal can be done with sheet metal snips, but thicker gauges and plates require shears or cutting torches. You can spend a few hundred dollars or a few thousand dollars on power metal cutting tools. Here are some of your options. Bimetal saw blades are the best for cutting metal, whether you are using them in your hacksaw or band saw. Make the blades in your hacksaw and portable band saw last longer not by forcing the cuts, but by allowing the saw to cut slowly and steadily. Excess pressure simply wears the blades out before their time. A horizontal metal-cutting band saw is a bench-mounted band saw with clamps to hold work pieces, and an automatic shut-off feature that turns off the saw when the cut is completed. The most common cutting capacity is 4 × 6 inches, which can cut through a rectangular piece of that dimension or a 4½ inch round. A portable metal-cutting band saw is slightly less expensive than the bench version and obviously more portable. The most common size cuts 4 inch stock. It is more difficult to create accurate cuts with the portable band saw, but the portability makes the machine more versatile in a do-it-yourself shop. A distinct advantage of the band saws is the very small kerf (less than 1/16 inch). Metal-cutting chop saws, cut-off machines, and angle grinders use abrasive wheels for cutting. Most of these produce a great deal of heat and metal dust, and cut pieces typically have burred edges. Metal-cutting circular saws with carbide blades produce cleaner cuts with less heat. They are also faster than chop saws. Both of these types of saws have larger kerfs (at least 3/32 inch)

Bending jigs can be made from many different items, including an engine flywheel (A), toilet flange (B), and various pipes (C).

than the band saw. A portable steel saw looks and operates like a standard circular saw, but is capable of cutting up to ¼ inch sheet metal, as well as cutting tubes and rods. Reciprocating saws can also be used for cutting rods and tubes. Manual or hydraulic metal shears and punches in small sizes suitable for home shops are relatively inexpensive. They are limited to thin-gauge materials, usually less than 16 gauge. And the heavier the material capacity is, the more expensive the machine. Manual snips are sufficient for cutting curves in sheet metal thinner than 18 gauge. Power nibblers make quick work of curves but are commonly available for 18 gauge or thicker.

Shaping

There are a number of ways to bend and shape mild steel. Brakes for creating angles in sheet metal, and ring rollers and scroll benders to create circles and scrolls are readily available. Unfortunately, the higher the capacity of these bending tools, the higher the price. You can make simple bending jigs from any round, rigid, strong form, such as pipe and salvaged flywheels, pulley wheels, or wheel rims. Using these jigs, it is fairly easy to bend a round rod up to ¼ inch and a flat bar up to ⅛ inch into complex shapes. A bench vise is handy for making acute bends. Tubing, rail caps, and channels can be bent with a heavy-duty electrical conduit bender. With practice and patience, you can create well-formed circles by making incremental bends with the conduit bender. Thicker materials are bent more easily if you have a long end for applying pressure; it is better to cut pieces to length after bending. Sliding a pipe over a short end will increase your leverage. When shaping metal into sharp bends, take into account the length needed for the radius of the bend. Cold-formed metal springs back somewhat, so it may take trial and error to find which size jig makes the bend size you desire.

Brazing

Brazing is very similar to soldering since flux is applied to tightly fitted metal parts that are then heated to the point where filler material will melt and be drawn into the joint. Silver soldering and hard soldering are terms incorrectly used to refer to brazing. Brazing is different from soldering because it takes place at temperatures over 840°F and below the melting point of the base metals. The metals are not fused, but held together by the filler metal adhering to the base metals through capillary action. There are a number of industrial brazing processes, such as dip brazing, furnace brazing, and induction brazing. The home welder is likely only to do torch brazing. Torch brazing can be done with an oxyfuel torch using acetylene as the fuel gas, or any of the other fuel gases. NOTE: Different fuel gases require different regulators, hoses, torches, and tips. For brazing to work, the gap between the parts must be between 0.002 and 0.010 inch. If the gap is too tight, the flux and filler will not flow evenly through the joint. If the gap is too big, the strength of the joint is lessened. Gaps between parts can be measured with a feeler tool, available at automotive stores. Items to be brazed must be absolutely clean and free from rust, corrosion, grease, oil, and cleaning compound residues.

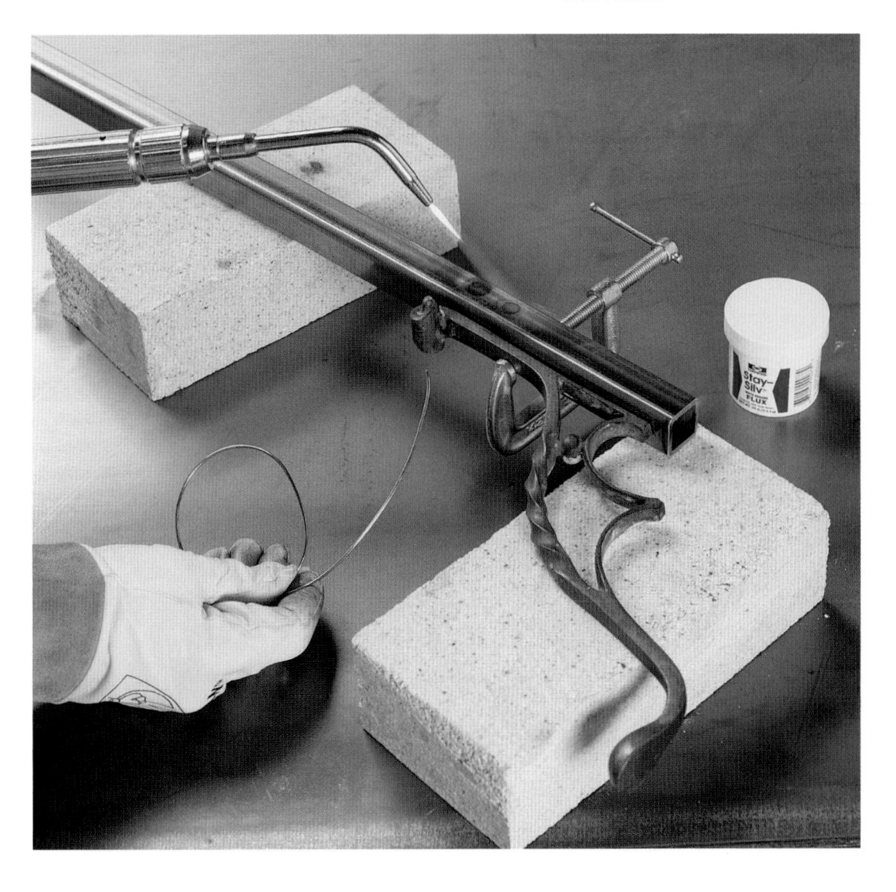

To braze, thoroughly clean and flux both sides of the joint area. Using a small torch tip, heat the entire joint area until the flux turns clear and starts to run. Add enough filler metal to fill the joint. After the metal has cooled, the flux residue can be removed with hot water.

Braze Welding

Braze welding is similar to standard oxyacetylene gas welding except the parent or base metals are not melted, so there is no molten puddle. Instead of a steel alloy filler rod, a flux-coated brass filler rod is used. Braze welding is often incorrectly referred to as brazing. Braze welding does not use capillary action to pull filler material into the joint—the filler

Braze welding is useful for joining thin metals to thicker metals. Heat both parts, directing more heat toward the thicker part. It may take a long time for the thicker metal to heat. Rest the metal on fire bricks to prevent the metal tabletop below from getting hot.

When both parts glow a dull cherry red, touch the flux-coated rod to the joint. The flux and the filler metal will melt. If the metal is molten or the fluxed rod comes in contact with the flame, the flux will burn and the filler metal will boil. This results in a poor joint in addition to giving off toxic fumes.

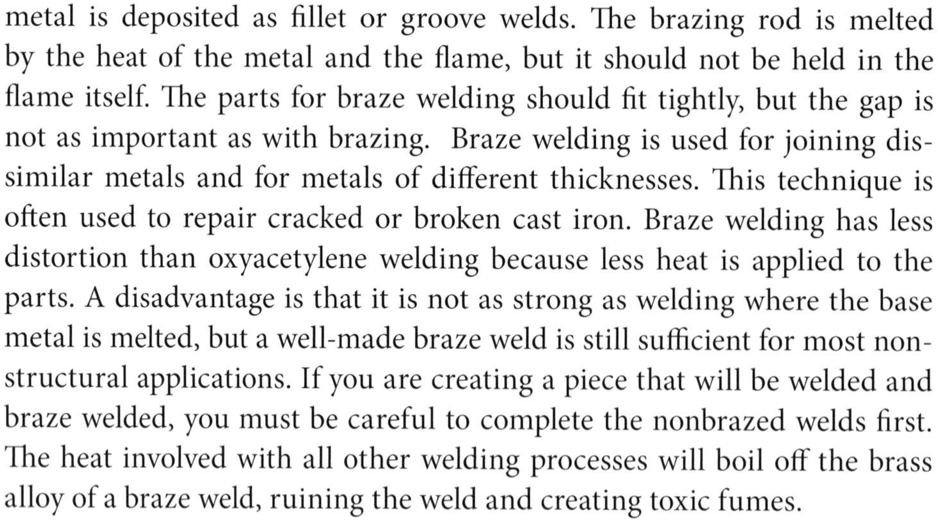

metal is deposited as fillet or groove welds. The brazing rod is melted by the heat of the metal and the flame, but it should not be held in the flame itself. The parts for braze welding should fit tightly, but the gap is not as important as with brazing. Braze welding is used for joining dissimilar metals and for metals of different thicknesses. This technique is often used to repair cracked or broken cast iron. Braze welding has less distortion than oxyacetylene welding because less heat is applied to the parts. A disadvantage is that it is not as strong as welding where the base metal is melted, but a well-made braze weld is still sufficient for most nonstructural applications. If you are creating a piece that will be welded and braze welded, you must be careful to complete the nonbrazed welds first. The heat involved with all other welding processes will boil off the brass alloy of a braze weld, ruining the weld and creating toxic fumes.

Shielded Metal Arc Welding (SMAW)

Arc or stick welding, or SMAW, involves the heating of the base metal to fusion or welding temperature by an electric arc that is created between a covered metal electrode and the base metal. The coating or covering on the electrode provides both flux and shielding gas for the weld. The electrodes come in 9- to 15-inch straight lengths in a range of wire thicknesses from $1/16$ to $3/8$ inch, hence the name "stick." SMAW is used extensively for fabrication, construction, and repair work because the machinery is inexpensive and

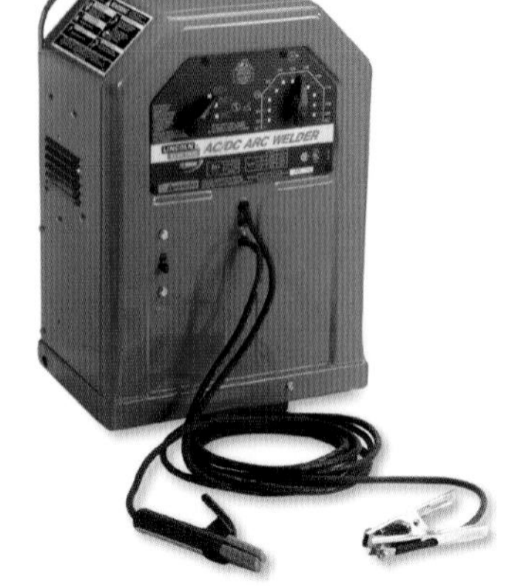

A shielded metal arc welder consists of a power source, electrode holder, and work clamp.

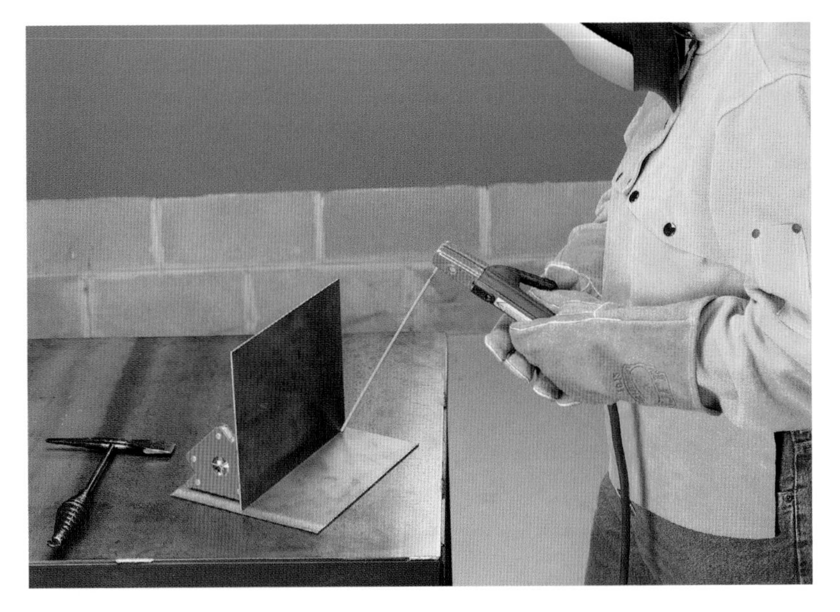

How to Weld with Shielded Metal Arc

Set up your material to be welded. Make sure the electrode holder is not touching the workpiece or worktable. Attach the work clamp to the table or workpiece. Turn on the machine. Adjust the range switch for the desired amperage. Wearing leather gloves, place an electrode in the electrode holder. Position the electrode over the area to be tacked, flip down your helmet, and tap or scratch the electrode on the area to be tacked to strike an arc. After making your tack welds, remove the electrode from the electrode holder. Check to see that the tacked pieces are aligned properly. If not, use a hammer to move them into alignment or break the tack welds and retack. Chip slag from tack welds so it does not contaminate the final weld.

Replace the electrode in the electrode holder, and position the electrode over the left side of the area to be welded. Hold the electrode at a 10° to 20° angle to the right. Flip down your hood and scratch or tap to strike an arc. The distance between the metal and electrode should not exceed the thickness of the electrode bare wire diameter. Move slowly to the right until the weld is completed.

To remove slag, hold the workpiece with pliers at an angle and scrape or knock the slag with the flat blade of the chipping hammer. Wear safety glasses when chipping slag.

PRECUTTING CHECKLIST

- Electrode and work leads are tightly attached to machine terminals.
- Cable insulation is free of cuts, damage, and wear.
- Electrode lead is tightly connected to electrode holder.
- Work lead is tightly connected to clamp.
- Work clamp is firmly clamped to worktable or workpiece.
- Floor is dry.
- Ventilation is adequate to draw fumes away from welder's face.

fairly simple, and the electrodes are inexpensive. There are drawbacks to SMAW: it doesn't work well on thin materials (less than ⅛ inch is difficult), electrodes need to be changed frequently as they are used up, and the protective slag coating must be chipped off each weld.

The SMAW process uses electricity, so there is always the possibility of receiving an electric shock or being electrocuted. When an electrode (stick) is placed into the electrode holder, it is "live." If the electrode touches anything that the work clamp is in contact with, the circuit will be completed and an arc will be struck. To prevent this from happening, always remove the electrode from the holder when you are not actively welding. Do not use your bare hands to insert or remove electrodes—always touch electrodes with dry gloves. Do not weld while standing on a wet or damp floor or ground, and do not weld outdoors in the rain. The electrode will be hot after use, so take care where you dispose of electrode stubs. A metal bucket is a good addition to your SMAW workshop.

The SMAW welding process produces ultraviolet and infrared rays, harmful fumes, and hot spatter. Protect your eyes with a #10 to #14 filter in a full-face welding helmet or hood. Heavy-duty leather welding gloves and a welding jacket with leather sleeves protect you from the molten spatter. Proper ventilation from an exhaust hood or fan is important, and it is a good idea to screen off your welding area so others are protected from the intense light of the arc.

The SMAW machines are available as either alternating current (AC) or direct current (DC) or with the capability of switching between the two. You often will hear SMAW machines referred to as "buzz boxes." The AC welding machines meet most home and small shop needs, are inexpensive, and are readily available. Because the alternating current cycles through a zero current between the positive and negative polarities, it can be difficult to strike and maintain an arc. The DC machines are easier to use and have many home and hobby applications, but they are more expensive. Because DC current can have its flow reversed, a DC machine has more versatility in terms of the electrodes that can be used. This versatility allows a wider range of welding positions, metal thicknesses, and metal types to be welded, making a DC shielded metal arc welder well worth the extra expense.

The equipment itself consists of the welding machine, which usually has one adjustment knob, an electrode lead with electrode holder (sometimes called a stinger), and a work lead with work clamp. You will often see work leads and work clamps referred to as ground leads and ground clamps. The work lead and clamp are not grounding the electricity; they are completing the circuit back to the machine.

The SMAW electrodes are solid, round, metal wires coated with flux and other components. In addition to producing shielding gas and flux, the covering may also contain additional metals for filler or alloying elements for the weld.

The American Welding Society (AWS) publishes standards for the electrodes. Electrodes come in diameters ranging from 1/16 to 3/8 inch in increments of 1/32 inch. The electrode diameter measures the wire itself, not the diameter of the wire with the covering. The electrode designation is inked onto the covering near the bare end of the electrode.

The diameter of the electrode will determine which amperage to use. Use the manufacturer's guidelines to determine amperage. The most commonly used electrodes are 6011, 6013, 7014, and 7018. The 6011 and 6013 electrodes will work with AC power; the 7000 electrodes will not. For a beginner, 6013 with DC power is generally the easiest electrode to use in terms of striking an arc and maintaining a consistent arc. Talk with your welding supplies dealer about what type of welder you are using and the materials you are working on to get the best electrode for your purposes.

The SMAW process produces a weld that is covered with a coating of ceramic-like slag. The flux and other components in the covering clean the material to be welded and also float out impurities in the weld. These impurities and flux solidify on top of the weld, which protects the cooling weld from the effects of oxygen in the atmosphere. The slag must be scraped or chipped away before the weld bead is covered with another weld layer or before the weld is painted or finished. Safety glasses should be worn during this procedure.

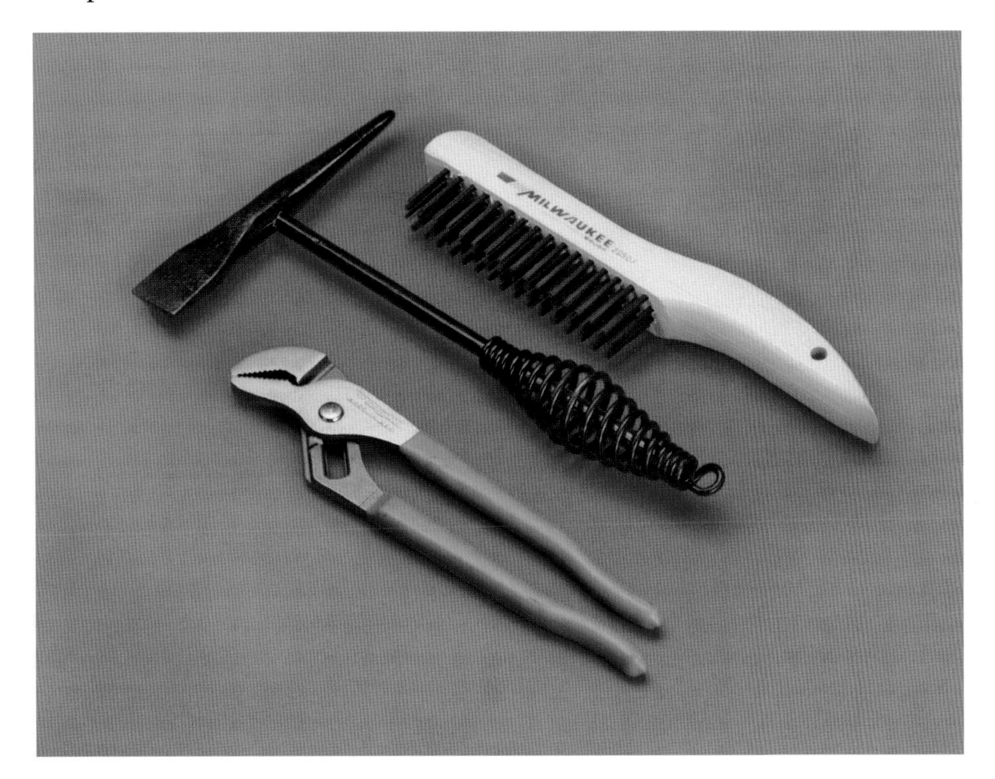

Useful tools for shielded metal arc welding are a pliers, chipping hammer, and wire brush.

Farm machinery parts are expensive. A well-organized storage system will help you avoid costly duplicate purchases because you can't find the piece or part you previously bought. Stocking your parts in separate bins will make finding them more convenient.

ESSENTIAL TOOLS

Adapted with permission from *How to Set Up Your Farm Workshop* by Rick Kubik (Voyageur Press, 2007)

Every farm, big or small, needs to be equipped to deal with a wide range of repair, maintenance, and fabrication jobs. Many farms maintain a workshop where tools can be kept and repairs can be made to equipment. Whether you have a small workshop or large one, orderly arrangement of tools and supplies will save time in locating the right tool for the job and makes for a more accessible work space.

Tool Organization

Tools can be organized for convenience of use and ready access. Your workshop can have its own system tailored to your tastes. One such shop can include a set of bins for new bolts and nuts in various sizes and a tray for holding "odds and ends" bolts and nuts left over from various disassembly jobs. There can also be a tray of leftover specialty fasteners that may find a later use.

Similar things are easier to organize when they come in large quantities, such as bolts, nails, or screws that come in boxes of ten or one hundred. This system of organization is easy to maintain when new items fit existing categories.

The system breaks down quickly, however, when you find yourself with many one-off situations where the similarity isn't as obvious, such as different size nuts and bolts that may be fine-thread or a metal other than steel. In practice, it's most likely to end up in the miscellaneous tray.

Try to maintain a clean workspace that has nuts, bolts, screws, tools, and small hand equipment in separate sections. A bit of organization along these lines will make it easy to find the right item for the job at hand.

Large pegboard racks can be an inexpensive solution for storing many of the items you use in your workshop.

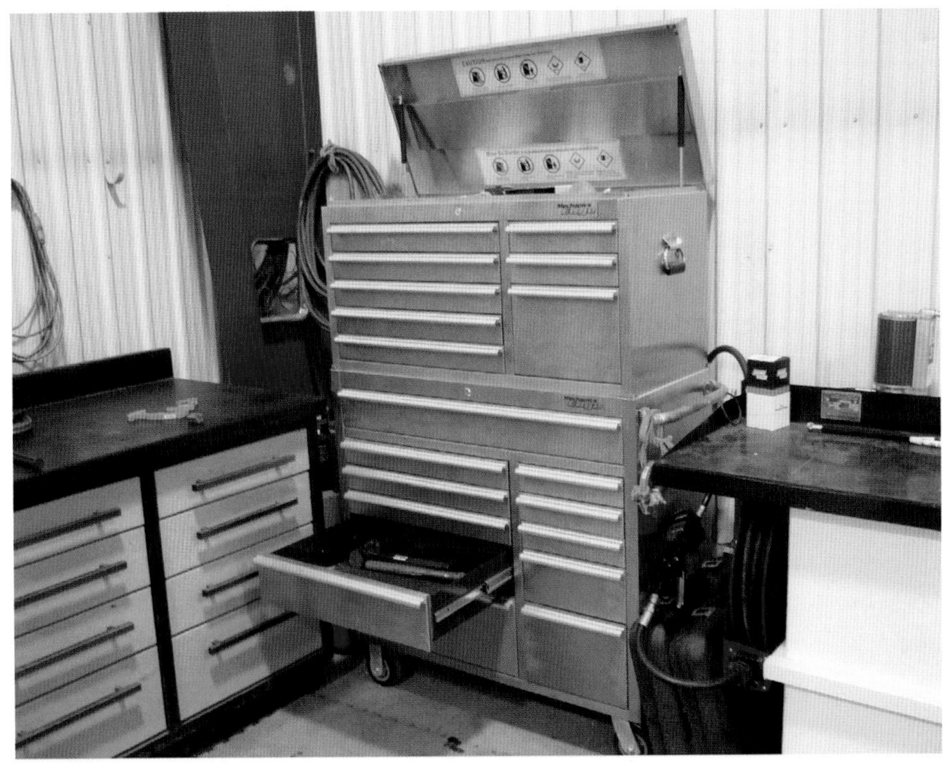

Take time to observe other workshops before setting up yours. Storage units can be expensive, so you should have some idea what will work best in your shop before making any purchase.

Toolboxes, Cabinets, Bins, and Racks

Organize your tools as one of the first things you do when starting to farm or building a shop. This organization can take the form of a portable box that opens to reveal many compartments, a big rolling cabinet, pegboard, or slot board mounted vertically on the wall, wall-mounted cabinets, cupboards under the workbench, and other custom-built solutions. Use cupboards and wall cabinets to use otherwise empty space under or above the workbench.

A well-designed workbench is one that keeps tools separate but still allows maximum use of the space involved. You can attach vises to this bench and have room to set stationary grinders. Sturdy benches make jobs safer as well.

Farm workshops need flexibility because of the different tasks that may need to be performed over the course of a typical week. These tasks may include welding, changing oil in different machines, tracking down an electrical fault in a tractor's instruments, calibrating planter units, rebuilding a carburetor, and many other variations.

No matter how good your tool storage is, it's not going to work for very long if things are never returned after use. As a rule of thumb, the easier it is to access the tool in the first place, the easier it is to put it back when you are done. If getting the right tool involved walking over to a fixed workbench rack, it's much less likely that you'll walk back to put it away until the job is definitely over, and by then, some other demand on your time may have arisen and cleaning never happens.

Workbenches

A key element in any farm shop is arranging tools and supplies for repair, assembly, fabrication, or modification. Your needs may include:

- A stable, well-lit, accessible platform to support and hold materials
- A convenient height in order to make your work more efficient and prevent repetitive motion strain
- Tool access that makes tools easy to find and put away
- Easy access to electrical outlets for power tools

An early example of a specialized workshop bench—the anvil. It combines surfaces for hammering, bending, punching, and tool-holding, and is mounted on a pedestal to make it accessible from any side.

The height of the work surface is a key component in whether the bench is comfortable and practical to use. The bench height should be at about wrist level for general work purposes. For precision work, you may need a bench that is higher so that you can see the fine details you work on. In such a case, you may choose to build a separate support box that can be laid on the main workbench to raise the work area.

Design your workbench before building it. A traditional approach in farm shops is to build 3-foot-high workbenches against the walls. The vertical surface at the rear of the bench may be utilized to hang tools from hooks or nails for at-a-glance storage.

QUICK TIPS OPTIONS FOR WORKBENCH DESIGN

- Make them smaller so they accumulate only the things you presently work on.
- Make them modular so that several can be clamped together if you need a larger work surface.
- Make them movable so workroom is available all around the bench, or can be moved next to where you happen to be working.
- Make them adjustable in height to accommodate various projects.
- Make them foldable with a clamping capability.
- Make them with built-in electrical power bars.

A creeper platform makes it easier to work under heavy equipment outdoors on gravel or a hard surface. They work well in shops with concrete floors.

Specialized Benches

In some tasks, such as welding, a heavy steel bench permits hammering and chipping and also improves the grounding conductivity of an electric welder. Because the bench is freestanding, you can move all around the bench to make welds where needed. This ability to keep your project stationary allows pieces to stay clamped in the vise or a tabletop jig until all welds are completed.

Sometimes work on larger machines, such as a tractor, can't be put on a bench. In this case, putting yourself on a special workbench or platform improves your ability to finish the work in comfort. A mechanic's rolling stool provides a comfortable position for working on pieces and equipment that would ordinarily be at about waist height and require frequent bending. For working under vehicles and some equipment, a mechanic's rolling creeper allows you to roll to the necessary places without lying on cold, wet, or dirty floors. Be sure to wear eye protection because dirt, rocks, and various fluids may drip down on to your face.

Lifts and Hoists

Many jobs in the farm workshop will involve lifting and moving heavy pieces of material, which are often awkward and involve twisting the body and extending the arms to get the piece out of a machine or back into position.

Tools that lift and hoist objects for you can often make moving items upward and sideways considerably easier and safer. Use a lift whenever it's practical, especially where you can reduce the risk of damage to parts and muscle strain or back injury for you or your assistants.

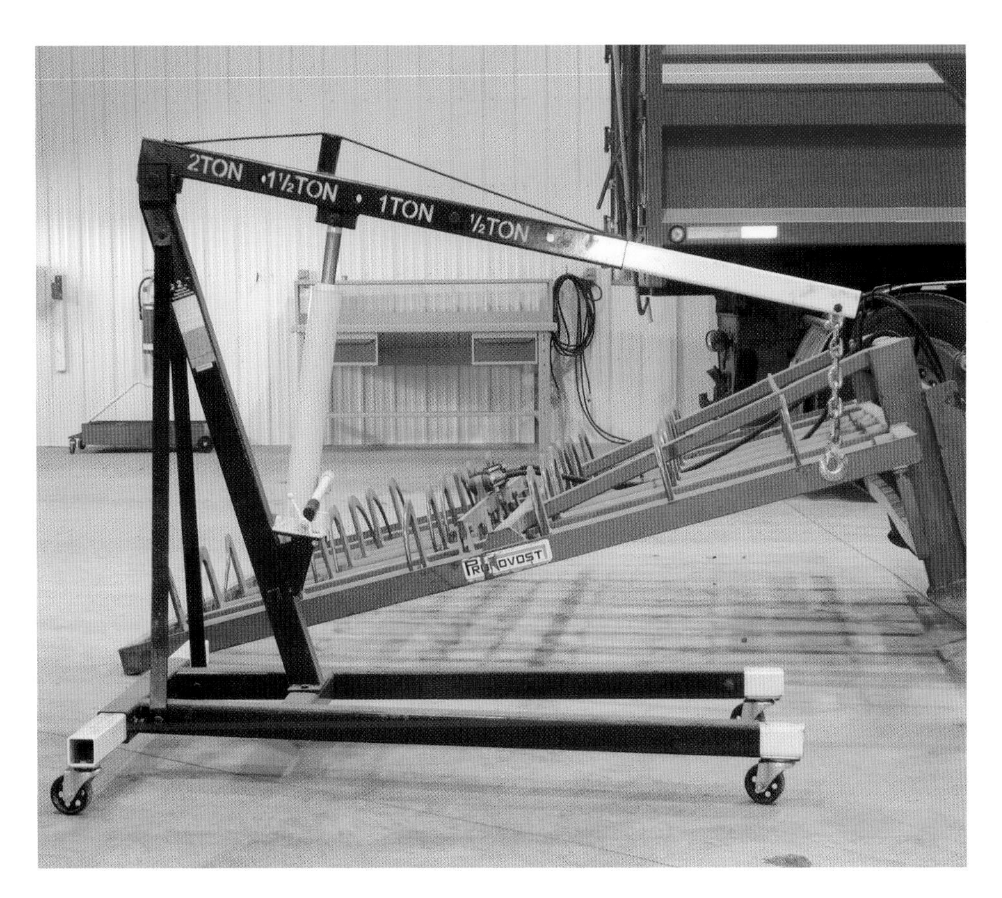

This specialized hoist is designed so that the lower support slips under the piece of equipment to bear the weight of the hoisting operation.

There is a difference between a lift or hoist and a winch when trying to move objects. A winch is meant to pull objects that do not lose contact with the ground or with some other friction-creating surface. Their purpose does not include pulling things up and holding them there in the absence of friction. The design of a winch is such that if you used it to lift a heavy object, free spooling could bring that heavy object crashing down with potentially disastrous results. The same sorts of problems occur with using a rope-and-pulley line. There is nothing to stop the load from falling rapidly if the rope slips or breaks.

A chain hoist is one of the most versatile lifting solutions. This multi-gear lift makes use of two chains: a smaller chain moves rapidly with light hand pressure and slowly raises or lowers the larger lifting chain. Ratcheting and electrically powered models are also available. Two valuable advantages of the chain hoist are that it requires very little effort, and when you stop lifting, the work hangs securely in place to make alignment easy during assembly.

Jacks and Supports

Machines do occasionally need to be partially lifted off the ground for work such as changing tires or servicing axles. Jacks and supports are needed for this kind of work. Jacks do the lifting, and supports hold the machine in place.

It is extremely important to understand that a separate mechanical support, such as wooden blocks, is necessary after lifting a piece of equipment. Do not rely solely on the jack alone to hold things up. Many farmers have

Your workshop needs an often-overlooked means to securely hold the equipment in the air while you are working underneath it. Stationary jacks such as these are lifesaving pieces that you should have in your shop.

been killed or injured because of a jack failure. Your shop should have plenty of wooden blocks around to suit whatever needs arise.

Presses

The arbor press uses a rack-and-pinion mechanism to push the moving shaft against the base. The rugged body casting keeps the whole thing from twisting when force is applied. Because hand pressure is all that's used with the arbor press, fine control of the shaft movement is possible.

The hydraulic shop press exerts much more massive amounts of force, typically in the range of 30 to 50 tons. The advantage of a hydraulic shop press for straightening metal is that heating is not required. When metal is heated and cooled, changes occur in its strength and hardness. By using a press to bend the metal back into shape, these changes are avoided and the metal stays at its original ductility and hardness specification.

A rivet press is a similar type of muscle-powered press used when replacing individual cutting teeth (sections) on sickle mowers, combines, or swathers. The new rivets are inserted and then the press pin is placed over them and tightened by using a wrench. As the rivet pin is tightened, the top of the rivet mushrooms out to lock it in place.

Compressed Air System

One of the key elements to include in your shop is a compressed air system to run a variety of tools such as wrenches, sanders, buffers, and drills. The same system will also provide compressed air for filling tires.

A generator and compressor that are attached to a cart are a useful approach to portability with a compressed air system. This allows you to take air power anywhere needed.

Air tools have several advantages including price and handling. They generally cost less to buy than electric-powered tools and are also lighter in the hand than electric tools. Air tools have a small turbine wheel that weighs less than an electric tool motor's armature and windings.

Your first step in installing a compressed air system is choosing a suitable air compressor and air tank. Base your decision on the air requirement of the largest air tool you have. Each air tool will have a specific average air consumption rating expressed as cubic feet per minute (cfm) at psi. For example, a rating of 4.5 cfm at 90 psi means this tool needs a supply of 4.5 cubic feet per minute of air compressed to 90 pounds per square inch in average working conditions.

Key issues when deciding which air compressor to buy also include life expectancy, frequency of use, relative noise level, and warranty. As a rule, better compressors will also have a longer warranty and up-rated frequency-of-use ratings.

Air lines are one consideration for your system. You can use flexible air hoses that can be stretched across the floor or to wherever you need them. Galvanized or copper plumbing pipe is sometimes used to run an air supply around the shop. These pipes are constructed like ordinary plumbing and positioned along walls where they can be accessed by the tool you are using. You need to make sure the joints are properly soldered to prevent air leakage.

One often overlooked danger is overfilling tires with air pressure. Over inflation of tires can cause the rubber to burst or make a split-rim band pop off the wheel. Always determine the correct inflation amount before filling the tire and do not exceed it. If you fill to the maximum pressure when the tire is cold, this is at the safety edge. Tires and the air inside usually heat up during operation, so air pressures may rise above safe levels.

Parts Cleaning and Degreasing

Getting the dust, mud, chaff, and manure cleaned off tractors and implements before they ever get into the shop makes the work more pleasant and removes contaminants that could make their way into engines and transmissions during repairs. Grease, sludge, and scale will need to be removed from engine and driveline parts such as driveshafts, oil pans, and cylinder heads in order to assess component condition and restore full performance capabilities.

A pressure washer is the first step in cleaning an item for subsequent operations in the workshop. Pressure washers use an engine and pump to boost water pressure and volume above what's supplied by the intake hose, as well as adding heat and chemicals to clean very quickly and thoroughly.

The pressure washer needs to be connected to a water source and electricity, so your workshop should have outdoor outlets to eliminate the need to string water hoses and electric extension cords long distances.

Sandblasting is a fast way to remove paint, corrosion, rust, and other hard contaminants from parts. Sandblasting works by using compressed

A pressure washer makes quick work of cleaning off machines before they bring mud, manure, dirt or straw into your farm workshop.

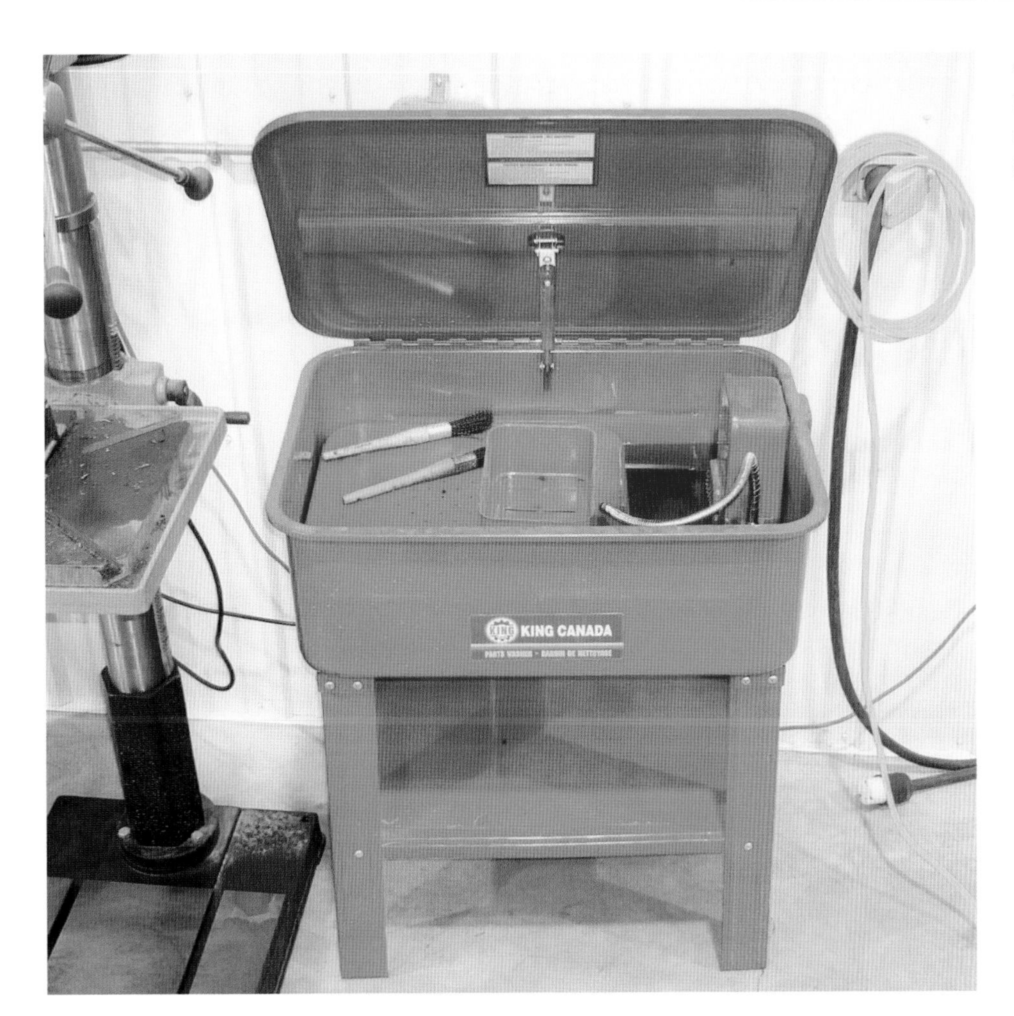

A large floor model solvent parts washer expands the range of parts that can be cleaned.

air to bombard a part with particles of abrasive materials. The abrasive materials most often used are ordinary sand, aluminum oxide, or glass beads. Due to the high volume of dust involved, sandblasters are used outside the shop. Pressure-feed or siphon-feed sandblasters require 80 to 125 psi of compressed air. Siphon-feed sandblasters operate by drawing abrasive material from an open container through a feed line. This type of sandblaster is less expensive and physically smaller, but one disadvantage is that it uses part of the compressed air energy to draw abrasive materials up to the outlet.

Solvent washing and degreasing occurs after the equipment has been cleaned off enough for closer work but before the oily or greasy parts have been cleaned. A stiff brush and a low flashpoint solvent may be enough for many small jobs.

Gasoline should never be used for parts washing, because the vapors can be too easily ignited. It was a common practice to use gasoline in earlier days, but not anymore. The ringing of your cell phone may be enough to spark a gasoline fire if there are a lot of fumes around.

In recent years many good water-based alternatives to the usual petroleum-derived volatile organic compound (VOC) solvents have become available. The VOCs are chemicals that evaporate easily at room temperature. They contain carbon and may or may not produce an odor. Everyday examples of VOCs include gasoline, low flash-point solvent, nail polish remover, chloroform, rubbing alcohol, butane, and propane.

Water-based solvents are usually less hazardous to the user than their petroleum-based counterparts. They clean by using a surfactant, such as soap or detergent, a corrosive or alkaline ingredient, or another type of chemical to remove soil from parts. Most aqueous solutions today are pH neutral so they are much less caustic to skin and surface finishes.

Precision Equipment

Certain specialized measuring tools may be helpful to include in your workshop. These tools are used in repair jobs or when determining if certain parts are still within their service limits. Depending on the extent of your farm equipment and your expertise, you may want to include the minimum equipment such as torque wrenches and coolant and electrical measuring tools. You may find others useful as well, such as micrometers, vernier calipers, and tune-up meters, but these tools will be used in specific situations and may not be needed often.

Torque wrenches provide a way to measure the amount of twisting force (torque) being applied to a bolt being tightened. Metal bolts are slightly elastic and stretch a measurable amount when tightened. That's why bolts in some critical applications such as cylinder heads and wheel bolts often have a specified tightening force. Too much or too little tightening may cause some sort of expensive and/or dangerous component failure.

Mechanical precision measuring tools such as the digital multimeter (left) are cheaper than they used to be. But don't overlook older technology like the analog multimeter (right). You may find these at auctions or even garage sales.

The load tester gives a quick indication of battery life. Don't leave it connected for more than a few seconds because of the enormous heat buildup in the tester during use.

You should refer to the service literature for specific information on bolt preparation before applying the specified torque. As part of the torque wrench equipment in your shop, provide a clean, moisture-proof storage of these precision instruments and avoid dropping them or placing other tools on them.

Tools for precise measurement of electrical system conditions are part of a well-equipped shop. These types of tools may include a multimeter or battery tester.

A multimeter measures power supplied (volts), resistance of flow of current (ohms), and current (amps). A multimeter can tell you exact voltage drops, resistance changes, or current flows at various points in the wire, or through an electrical component. The test light only gives a "yes or no" indication of voltage.

A multimeter will help narrow down the area that's creating a problem. For example, if a tractor's starter or any of the instruments fail to operate properly, it's easy to use the voltage (V) measurement function to determine if there is any difference in voltage at one end of a wire and the other. If there is, checking voltage at various points along the wire, such as between the wire and connector or connector and ground, can quickly pinpoint the fault. Different types of multimeters are available including digital and analog (dial type), which show a range selection.

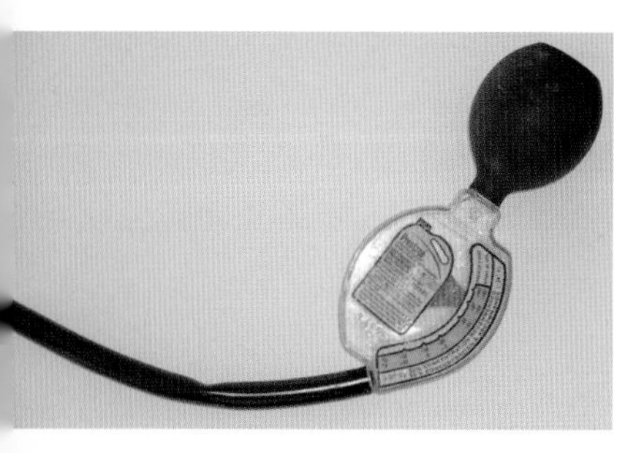

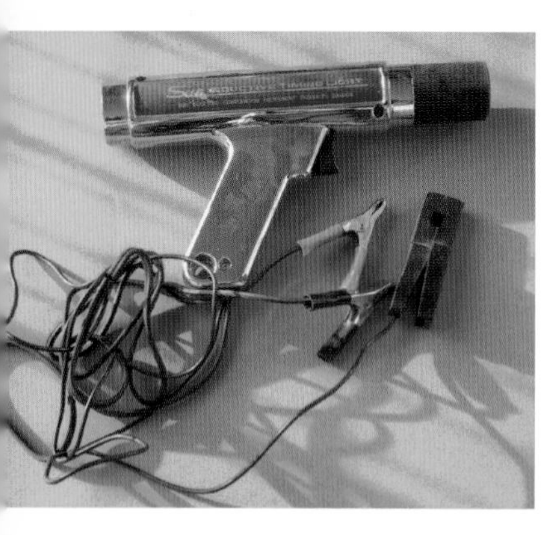

TOP: The coolant hydrometer works on the same principle as the battery hydrometer but should only be used for measuring coolant concentration. The readings of two different kinds of hydrometers are of use only with their respective systems.

BOTTOM: Tune-up meters are useful for diagnosing problems with the non-computerized and mechanical ignition systems in older engines.

Battery testing uses an electrical measurement tool called a battery hydrometer. This measures the specific gravity of the electrolyte in batteries, which indicates how much charge remains in the battery. A battery hydrometer and coolant hydrometer are different instruments and one cannot be used in place of the other.

The battery load tester verifies that the battery can reliably deliver its specified power. The load tester is always useful in determining whether it's worth recharging a battery or whether it's time for a replacement. With a tester, you do not need to remove a single battery or a set of batteries to test them.

An ignition diagnostic meter and strobe-type timing light are necessary parts for working on older gasoline engines that have point-and-condenser ignitions and variable timing. These multiple functions of the diagnostic meter are key to making several adjustments that will keep the engine starting and running smoothly. The diagnostic meter allows precise indications of how long the ignition points stay open. The tachometer function makes precise tuning of carburetor idle and low-speed and high-speed fuel mixture screws easier. The timing light permits a precise setting of when the spark plugs fire and is used in conjunction with the diagnostic meter's tachometer. It also permits a check of whether the distributor's spark advance mechanism is functioning correctly. The multifunction diagnostic meters and strobe timing lights made for automotive use are also capable of handling tune-up task for gas-engined tractors and other farm equipment.

Precision measurement tools should be stored in a clean, dry place separate from other tools in order to avoid damage.

Grinders

Abrasive-wheel grinders, both permanently mounted and portable handheld types, will find plenty of use in shaping metal, beveling edges to prepare for welding, removing burrs, and removing the small annealed zone that results from metal being cut with a torch. Also, the wire wheel that can be permanently mounted on one end of a stationary grinder is very useful for removing paint and corrosion on metal.

For a stationary grinder, mounting on a freestanding heavy base or pedestal provides maximum freedom to move work pieces around a grinding wheel. Grinders can also be mounted on the bench top, but this may restrict movement of the work pieces.

Grinders come in a variety of sizes and models. Ratings may be in terms of wheel size or output power, which is the power available for tool operation after heat loss. Handheld portable grinders are lightweight and can be easily moved from point to point.

With any type of grinder, look for safety features such as sturdy, easy-to-adjust work piece rests, grinding wheel guards, and spark shields.

A small vise on the table of the drill press keeps the work steady during drilling, especially in the critical last part of the cut.

Drills and Vises

A stationary drill press (also known as a pillar or pedestal drill) or a handheld drill is a vital part of your workshop's equipment. These tools help to make holes in metal for fabrication or repair. Stationary drills are more precise because they don't move; a handheld drill may shift while pressure is being applied.

The use of a drill often requires a vise to hold the metal piece being worked on. The high-torque drilling that makes them so useful also contributes to one very common safety problem: the work piece being jerked out of the operator's hands if the bit jams in the hole. This is most common when the operator just wants to quickly drill a hole and uses a lot of pressure without clamping down the work piece. The easiest solution is to always clamp the work piece in a vise to hold it while drilling. Some vises simply clamp the work, some allow you to tilt the work piece to make angled holes, and some have calibration capabilities.

Vises have many uses in a farm shop. They hold metal pieces stationary to be worked on whether you are welding, cutting, grinding, drilling, repairing, metal sawing, bolt tightening or loosening, and a vast array of other jobs. They can be portable, stationary, large, or small.

Files

Files are used for shaving, smoothing, and fitting metal parts, and for basic sharpening, such as with axes and lawnmower blades. For farm workshop metalwork, a set of flat, round, and triangular double-cut (grooves running both ways) machine files and single-cut (grooves running only one way)

Metal work involves applying considerable force to the work piece, so equip your shop with a strong vise to hold the work.

mill files will be useful. Mill files leave a smoother finish than machine files but cut more slowly. Files are made in four levels of coarseness: coarse, bastard, second, and smooth. Again, the finer cuts are slower. File sets often come with a mower file with a handle on the end. These files are used for sharpening hay mower blades and may be used for many other jobs. Needle files are smaller, which make them useful for cleanout work in tight places. The auger bit file is good for filing right against an edge you don't want disturbed. A roll of emery cloth (abrasive on a strong cloth tape) allows filing and polishing round shafts or holes to remove rust or corrosion without disturbing the diameter.

Hammers

Your workshop should be equipped with several sizes of metalworking hammers, which differ quite a bit in shape, weight, and striking face hardness from the claw hammers that are used in carpentry. Carpentry hammers are made for striking nails made out of metal softer than the hammer's head,

Body workers' hammers and dollies are available in inexpensive sets and help you work out wrinkles and bends in sheet metal such as power-take-off shields.

and metalworking hammers are made to handle harder metals. Carpentry hammers should not be used for metalworking unless you have nothing else on hand.

The ball-peen hammer, also known as an engineer's or machinist's hammer, has a flat striking face on one end and a round striking face on the other. The round face imparts a rounded shape to the metal being formed, while the flat face is useful for all types of metal striking work. The original function of the round face was to peen (mushroom) protruding metal, such as rivets on cutter-bar sections or the ends of roller chain connector links. Hammers are graded by the weight of the head. They can be bought individually as needed or in sets including 8-, 12-, 16-, 24-, and 32-ounce sizes, which should take care of the needs in any farm workshop.

Other useful metalworking hammers include the cross-peen, clubbing, dead blow, soft face, and slide types. The cross-peen hammer has one flat face and one face that tapers perpendicular to the handle. The pointed face is used for hammering straight-line bends in metal, expanding the ends of rivets in an X-pattern, or driving metal pieces sandwiched between other flat pieces.

The clubbing hammer has a heavy head and short handle, and the flat faces are larger than those of the ball-peen or cross-peen types. The larger size of the flat faces makes it easier to strike punches and chisels without missing, while the heavy head and short handle allow you to impart lots of striking force using a slower, shorter, easier-to-control swing.

LEFT: Whether you prefer plastic or metal, it's easy to get good-quality calibrated jugs to eliminate measurement guesswork.

RIGHT: This funnel looks something like out of the steam age, but it's still a very useful piece of equipment because it bends to fit the area in which fluids may need to be poured. Having a screen in the outlet will keep contaminants from flowing down into the reservoir.

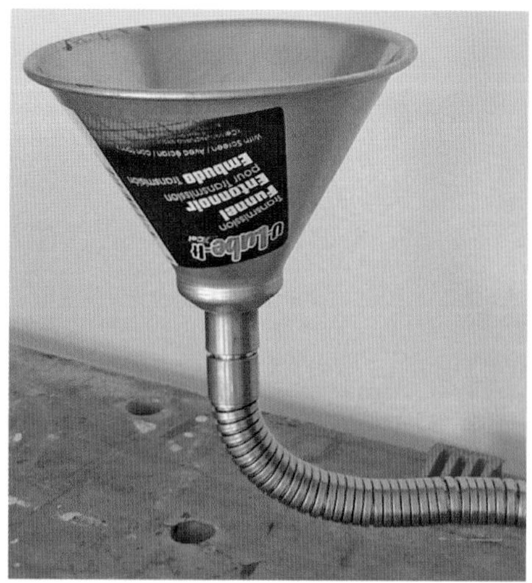

The dead blow hammer is designed so that it does not bounce after an object is struck. This trait is useful if you have to strike a very hard shaft in order to loosen it. The soft-face hammer allows you to apply striking force to metal that might suffer damage from an ordinary hard hammer face, such as loosening a cylinder head that is stuck to the block.

The slide hammer can be used for removing parts that are stuck together tightly, such as bearings and the casting that steadies them. One end is hooked to the part and then the weight is slid rapidly along the shaft until it strikes the stop and jerks the whole assembly outward.

Measures and Filling Tools

Efficient service and long life of machines depend on a sufficient source of the right fluid, whether that is a lubricant, hydraulic/brake fluid, coolant, or some other type of fluid such as windshield washer fluid.

Many of the fluid fill holes on tractors, implements, trucks, and other equipment are quite often in awkward-to-reach, hard-to-see locations. Having the right measure and fill container will help with any job.

When it's time to fill a fluid reservoir back up, a few inexpensive pieces of shop equipment are handy unless the reservoir is designed to hold a whole number of quarts, gallons, or liters. Properly calibrated measuring jugs are readily available and they eliminate guesswork and estimation involved with makeshift containers.

Measuring jugs need to have quantity markings clearly stamped into or molded onto the sides of the vessel. For plastic containers, the clearer the body material of a measuring container is, the easier the fluid level can be seen as it fills up with fluids that are dark enough to provide good contrast in low light. When filling the jug, provide a flat, level surface so that you can be certain the level is correct.

Look for a fluid measuring jug with a well-defined pouring spout so that fluid doesn't dribble down the sides when pouring out. A completely enclosed spout will pour slower but more accurately than a spout with a partly or fully open top. Flexile spouts allow you to work in close places.

A grease gun is an indispensable filling tool that you will use often. Several types are available, including hand-pump and electric, that make greasing any machine easy. The easier it is to grease the machine, the more likely it will get done properly.

You will encounter many needs for different tools both for construction and repair. Always keep your safety first in mind when purchasing any tool to use in your farmshop.

Chapter 5

Farm Safety

The farm environment has many hazards, and accidents can happen quickly. The danger is compounded by the fact that you're often working alone and far from medical expertise or emergency assistance.

This chapter is adapted with permission from *The Farm Safety Handbook* by Rick Kubik (Voyageur Press, 2006)

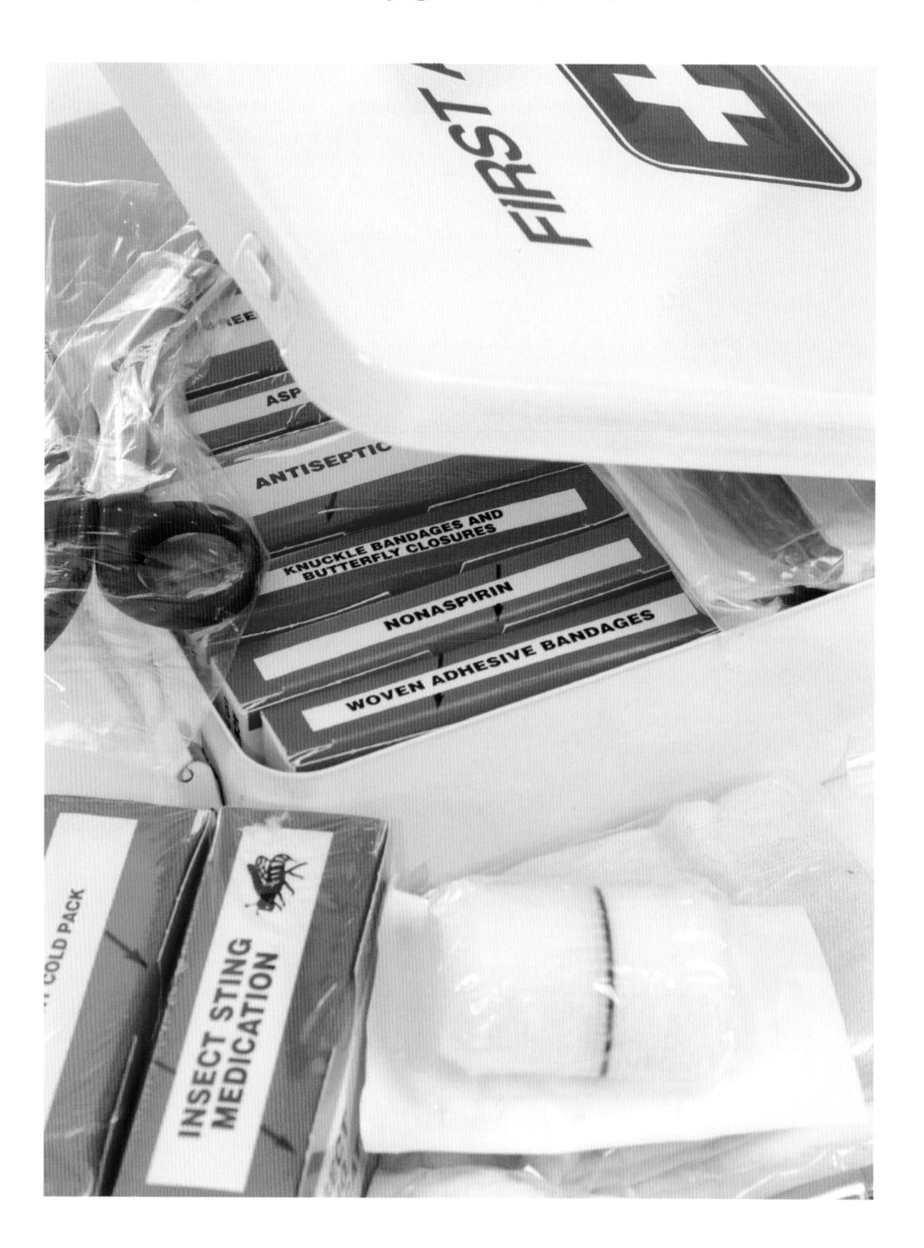

Inspect your first aid kits every year and make sure they're easily accessible.

Educate yourself and your family about how to handle machinery and animals safely and how to prepare for emergencies. Take time to learn about hazards you're likely to encounter and consider the best way to respond. This knowledge will increase your chances of living a long and healthy life on your farm.

BASIC SAFETY TOOLS

Start all your jobs with the right tools—including tools to keep you safe. Below is a list of basic safety equipment. Remember that just having the tools isn't enough: they're of no use to you if they're lost or broken.

- Operator manuals. These guides discuss safe operating methods. Keep them where they are easy to find.
- Contact information for emergency services. Post this prominently near every phone or have it easily available on your cell phone.
- First aid kits. Inspect first aid kits at least every year and make sure they contain sufficient, up-to-date supplies. Keep kits where an injured person can quickly and easily get to them.
- Fire extinguishers. Inspect them at least every year and recharge as needed.
- Flashlights. Check them regularly to make sure they'll work when you need them.
- Water. Locate a reliable source of water on your property for fighting larger fires.
- Emergency plan. Put together a plan of where to go and whom to contact in an emergency. In particular, teach children what they need to do.
- Learn the right way to manage machinery and livestock so you can put that knowledge to use to protect yourself and your family.
- When you buy, rent, or borrow equipment, get training that shows you proper and safe usage if you are unsure.
- Read and understand operator manuals.
- Read and understand warning labels on the machine. They alert you to potential dangers of unsafe operation.
- If handling livestock, learn safe handling techniques and strategies.
- Make time for maintenance. Accidents can occur because of machines that are not repaired or maintained correctly. Keep driveline guards in place and properly secured.
- Learn the signs of danger. Some products used on farms are clearly marked with unique symbols to alert users of potential hazards.

Safe Towing

Solving towing situations is not simply a matter of tugging with a towing line. Parts such as ball hitches, clevises, chains, and complete bumpers can break loose and become dangerous missiles capable of creating great danger for those nearby. All towing materials—chains, cables, tow straps, and nylon ropes, along with the hooks at their ends—are dangerous when they recoil from a stretched position.

When towing, hitch to the tractor draw bar, not to other parts of the tractor, and make sure you use a tow line big enough to handle the weight. If you are using a fabric strap as part of the tow line, it should be connected so that if any type of failure occurs while the strap is under tension, the stored energy in the strap will not cause objects to be thrown or propelled toward workers.

When towing with a tractor, hitch only to the drawbar. This ensures the applied load will keep the tractor level front to back. Do not attach a tow line to any tractor part above the drawbar. This can result in a backward overturn.

During transportation of an auger, elevator, or other high-profile equipment, contact with overhead wires is a risk. Scout your transport area and route and lower the equipment to transport position whenever possible.

Take up slack slowly and exert a steady pull. Ask the operator of any stuck machine or vehicle to drive the same direction you pull. Taking a running start to try to jerk the load free can stress any tow line to the breaking point or cause a tipover.

Overhead Electrical Wires

Electrical service wires in many rural areas and farmyards are strung overhead from pole to pole, rather than underground. Normally, the lines are out of the way and tend not to be noticed, but you need to watch carefully for overhead wires when moving any tall equipment, such as a grain augur, conveyor, fold-up implement, or a tractor with a raised front-end loader, or before raising long pieces of metal, such as a ladder or pipe.

Before attempting to repair electric lines broken by farm equipment, move the transfer switch securely to the OFF position.

Stay away from any downed power lines during or after a wind or rain storm. Contact your local utility emergency services immediately.

Ladders

Ladders are one of the biggest safety hazards encountered by most people in their day-to-day work and home life, and the hazards are just as common on the farm. Make sure any access ladder, such as to a grain bin or building roof, is securely placed to prevent both falls and the risk of being stranded atop a building if the ladder falls down.

To keep children from scaling buildings with permanent access ladders, keep the end of the ladder above a child's reach and instruct them not to climb any ladder without your assistance.

Yard and Workshop

You can avoid many hazards in a typical farmyard and workshop with basic practical precautions. Minimize children's horseplay around equipment and high storage units such as fuel tanks.

Install handrails on all elevated access platforms to prevent falls from accidental steps backward or slips while turning. Install solid steps with handrails in elevated areas where you may carry feed pails, bales, or heavy items.

If you are using irrigation equipment where the valves are electrically controlled, keep both electrical and water connections in good repair with no leakage. Avoid being near a site where water and electricity can mix.

Other Potential Hazards

Other examples of potential hazards and solutions include the following:

- Hazards can occur when accidentally backing into or running over pipes with trucks and implements. Put simple barriers around hazardous sites to help prevent accidents.

Secure access ladders to prevent falls and reduce the risk of being stranded atop a building, should the ladder fall down. Chain or tie ladders to a sturdy permanent structure before ascending to do any work.

- Careless removal of a radiator pressure cap can blow steam and coolant into your face and eyes. Let the engine cool down completely and turn the cap slowly to release pressure.
- In the workshop, do not rely on jacks alone to support a weight. Use solid stands to block the weight from falling.

A wide variety of potential hazards awaits farmers, and these are only a few. Use common sense, anticipate potential hazard sites, and stay alert while attending to problems. You will help lower the risks for injury or death to you and those around you.

TRACTOR SAFETY

Rollovers are the single deadliest type of injury on farms; runover accidents caused by bypass starting or falling off a moving tractor cause the greatest number of tractor-related incidents.

Some of the same design factors needed to make tractors useful in normal situations make them dangerous if they are not operated with due care.

Minimize your risks by understanding tractor stability, remaining alert while operating a moving tractor, and installing rollover protection.

Tractor stability involves several dynamics, including design, weight, wheelbase, area of operation, and road-speed capabilities. The design, weight, and wheelbase are set by the manufacturing company as part of their tractor model construction. When you purchase a tractor, take these dynamics into account in relation to the terrain on your farm. Some general considerations should include the tractor's center of gravity, width of front end wheel configuration, tire condition and tread, and whether you will attach a front loader.

The center of gravity of your tractor can change relative to the terrain it drives over. This issue is typically not a concern on level ground as the weight settles directly over the area on which the tractor is moving. The tractor's balance will shift on side slopes or side hills, when climbing a slope or hill, or when descending a slope or hill.

You can mitigate the pull of gravity on the balance of your tractor by understanding the seven danger zones:

1. Side slopes or hills
    - Tall grass can hide potholes and large rocks, so walk the area first to check for hazards.
    - Be sure the tractor's brake pedals are locked together and test the brakes before going on the slope. If only the uphill-side brake is applied on a side slope, the tractor will slew upward and increase the risk of a sideways rollover.
    - Wet grass or loose soil on the slope can lead to the tractor sliding sideways, then tipping sideways if the tires stop sliding.
    - Driving the tractor near the top of a slope may put it on less stable soil. If the soil gives way, the tractor can start sliding down and tip over sideways.

In situations where tipping may be a concern, such as working on a slope, side-to-side stability can be improved by setting the wheels as wide as possible. Wheels may be able to move farther apart on the axle, or in some cases they can be reversed.

Front axles on many tractors have some provision for adjusting width. To improve stability, use the maximum width practical for your job.

- If it's practical, work up and down the slope or side hill rather than across it. Travel down the slope going forward; travel up the slope in reverse.
- If work on side slopes is continually necessary, set the tractor wheels. For increased stability, use a low-profile tractor or install smaller-diameter wheels on an existing tractor to lower the center of gravity.

2. Climbing a slope
   - Climb with the heavy end up the slope (in reverse).
   - Wet grass or loose soil can lead to the tractor spinning the rear wheels, then flipping backward if the tires suddenly find good traction.

Traversing a slope can cause a tractor to tip sideways. Factors that increase the danger of sudden tipping include a downhill wheel falling into a pothole or animal burrow, an uphill wheel jarred upward by a stone or dirt pile, or a lifted implement that raises the overall center of gravity.

- Up-slopes often have a very steep (sometimes near vertical) final few feet at the top. This sudden increase in slope increases the tipping hazard.
- If the tractor starts rolling backward, avoid sudden application of the brakes because the nose tilting up dramatically increases the risk of a backward rollover.

3. Descending the slope
   - Walk the path first to check for potholes or large rocks that may be hidden.
   - Descend with the heavy end up the slope (travel front end forward, rear wheels back).
   - Travel straight down the slope, not at an angle, to avoid risk of a sideways roll.
   - Use both brakes in tandem rather than only one at a time to decrease risk of the tractor sliding sideways.
   - If traveling down with a load in the tractor's front-end loader, keep the bucket close to the ground to lower the center of gravity.
   - If you can only get off the road and out of the field by traveling down a steep ditch bank, consider either building a less-stress ramp or traveling to and from the field in the ditch bottom and avoid slopes altogether.

Climbing a slope in a forward direction can cause the tractor to tip backward. Factors that increase the danger of tipping include the uphill axle getting jarred upward by rough ground or applying the brakes when the tractor starts rolling backward.

4. Sharp turns
   - Before traveling above field speeds (3.5 to 4 miles per hour) be sure the tractor's brake pedals are locked together for use rather than using just one brake if needed.
   - Avoid sharp turns at road speeds or on slopes and side hills.
   - Watch for ruts or potholes that could jerk the steering wheel out of your hands and lead to an unexpected sharp turn.

5. Starting off with heavy loads
   - Hitch heavy loads only to the drawbar.
   - Rollover protection structures (ROPS) should not be used as a hitching point for a chain, rope, or cable.
   - Take up any slack in load connection slowly and smoothly.
   - Stay ready to cut power immediately if the front end of the tractor starts to rise.
   - Do not pull at an angle because it adds the risk of a sideways rollover.

6. Road speeds
   - Pick a safe travel path and stick to it. If you have to pull over to let traffic pass, look for a safe pullout or wider spot in the road, or slow down or stop until clear.
   - If the road has heavy high-speed traffic, consider traveling to and from the field in the ditch bottom or ends of fields or find an alternative route altogether.
   - Avoid sharp turns at road speed.

- Slow down for big bumps in the road to avoid jolting the tractor steering.
- Remain alert while on the road at any speed.

7. Edges of waterways
   - Test the ground by walking it first. Soil close to the edge of a natural or man-made waterway is likely to be soft and have and reduced load-bearing capability. If the soil suddenly gives way, this increases the risk of the tractor tipping down the bank.
   - Avoid sharp turns near the edges so a tire doesn't suddenly slip over the edge of the bank.
   - Keep the steering wheel steady when you turn to look back at any equipment to avoid turning towards the bank.
   - Work at a reduced speed along banks.
   - If the equipment being pulled projects to one side of the tractor, make your initial pass with the implement projecting toward the watercourse. This positions the tractor away from the edge.

Rollover Protection Structures (ROPS)

Tractor manufacturers add ROPS, which are roll bars or roll cages designed for farm tractors and can be installed if they do not exist on your tractor. The ROPSs absorb the impact energy during a rollover to create a zone of protection for the operator. However, for them to work effectively the operator must stay within the protective zone by using a seat belt. If the tractor rolls over and the operator falls out of the seat or tries to jump clear, the ROPS cannot do their job.

Working too close to the edge of a stream, river, canal, or other waterway increases the risk of a sideways rollover and adds the danger of drowning in the water or mud at the bottom of the slope.

The Occupational Safety and Health Act (OSHA) requires an approved rollover protection structure (ROPS) for all agricultural tractors over 20 horsepower that were manufactured after 1976 and are operated by a hired employee.

Runovers when starting from the ground can be prevented with an inexpensive remote starter switch. The switch wires are clipped to the starter solenoid terminals.

Runovers

Modern tractors have many built-in safety features, but many older ones do not. Older tractors have a variety of starting mechanisms, and these can cause a risk if not used properly.

Runovers most often occur when the operator doesn't observe other people around the tractors when they start movement or try to start the tractor while standing on the ground, often in front of the rear wheels. It usually occurs when a shortcut is taken that bypasses all safety considerations. On tractors with bad wiring, broken starter switches, or weak batteries or starters, some operators are tempted to work around these situations by jump-starting or short-circuiting part of the starter wiring.

Runover accidents often occur as people assist their partners in hitching equipment to tractors. Inadvertent movement of the tractor can easily injure a person in the close quarters between tractor and implement. Take extra precautions when someone is helping.

FRONT-END LOADERS

A front-end loader is often one of the first optional attachments installed on a farm tractor and makes the tractor capable of lifting and transporting objects and material around the farm. However, loads in the bucket always affect the stability characteristics of the tractor. For safe operation and reduced stress on the loader components, the operator should be aware of the changes and how to minimize their effects.

Bucket Loads Affect Stability

Viewed from the side, a tractor with a front-end loader behaves like a lever and fulcrum system. At one end of the lever is the load in the bucket; the weight of the tractor is the load at the other end. The fulcrum, or pivot point, is the front axle of the tractor, which is why a load in the bucket tends to lift the rear wheels. If the tractor is moving, rear-end bounciness also increases if the tractor goes down a slope or if the wheels hit a bump or hole.

One critical consequence of this levering action is that lightening the load on the rear wheels of the tractor also reduces the effectiveness of the brakes. Farm wheels only have brakes on the rear wheels. If a heavy load raises the rear wheels, no braking power is available. The tractor may roll freely forward or backward, depending on the slope of the ground.

Viewed from the front or back the front-end loader has another type of lever effect. The pivot point is where the tires touch the ground, and the force at the end of the lever load is the load in the bucket. The higher the bucket is raised, the longer the lever, so the more tipping force the tractor receives if the load moves away from straight above the middle of the tractor.

In addition, having the front wheels set close together increases the effect of any sideways tipping force, while wheels set farther apart makes the loader-equipped tractor more stable. The stability benefits of wider axle settings are also the reason to avoid using front-end loaders on tricycle-type tractors.

A tractor with a front-end loader is subject to both sideways and front-to-back tipping forces. Turning increases the effects, and the faster the travel speed, the more pronounced the effect.

The final stability point of view to be considered is at the bucket itself. If the loader arms are raised very high without adjusting the forward-and-back angle of the bucket, the risk of the load in the bucket rolling back toward the operator is increased. Many accidents of this nature occur when handling large round bales in an ordinary front-end loader.

QUICK TIPS FOR OPERATORS

- Insist that there be no riders unless your tractor is equipped with an approved trainer seat inside the ROPS safety zone.
- Do not let riders in any front-end loading bucket, the platform, steps, fenders, hood, or standing on the drawbar.
- Before moving any tractor, visually and audibly confirm the location of anyone near and don't move until they are out of your path.

The higher the load, the more forward and sideways leverage is exerted on the tractor. A tractor can roll over from this position if it's on a slight side slope or if the wheel hits a bump or hole. Keep your load low to the ground during any transport.

If the rear end of the tractor starts to life as the loader is raised, or if it feels bouncy as the tractor moves, there is too much weight in the front end and/or too little weight at the rear of the tractor. Stop lifting, adjust the load, and add ballast to the rear.

Danger Zones and How to Deal with Them

Working with tractors presents five major danger zones. Here is a list of the zones and suggestion on how to safely avoid or work around them.

1. Heavy loads
   - Consult the manual to find out the recommended maximum bucket load.
   - Fill the bucket less when handling very heavy materials, such as damp sand.

- Add weight to the rear of the tractor with wheel weights or by attaching a ballast box or heavy implement to the three-point hitch.
- Lift the bucket slowly to prevent sudden tractor instability.
- Keep the bucket low when traveling.
- Reduce travel speed on slopes, in turns, and on rough ground.

2. High lifts
   - Use high lift on flat ground.
   - Raise the load slowly to prevent sudden tractor instability.
   - For loads that can roll or shift, such as hay bales, use only loaders equipped with a grapple that prevents backward rolling.
   - Lower the load for traveling.
   - As you lift the loader arms, adjust the angle of the bucket so that the floor of the bucket does not tip too far backward. Self-leveling loaders automatically keep the bucket level.
   - Watch out for overhead electrical wires and other obstacles when the loader is raised.

3. Field work
   - Remove the loader if feasible. Many loaders have a "quick detach" design that allows easy removal and attachment without tools.
   - If the loader is not removed, remove the bucket.
   - If possible, lock out the loader controls.

4. Riders in the bucket
   - Although it looks like a convenient place to ride, especially for children, the front-end loader is not a suitable people mover. Inadvertent

Sometimes it may be necessary to have the loader arms elevated to work on the tractor. Be sure to block the loader arms before working on it. Unexpected downward movement of the arms could trap, injure, or kill.

movement of the loader controls could put riders in danger of having their legs drag on the ground, especially at road speed. But an even worse danger is that the rider could be bounced out of the bucket if the tractor hits a bump in the road.

5. Working around front-end loaders
   - When working on or around front-end loaders, block the loader with support stands or solid blocks of wood before putting any part of your body in a position where it could be crushed by the loader arms. In older tractors with worn hydraulic seals and pumps, hydraulic leakage may cause the loader arms or bucket to creep unexpectedly.

HITCHES

The area between a tractor and the implement can be a very dangerous spot for the person doing the connection. Typical injuries in pin connections involve relatively minor scrapes, bruises, or muscle strain, but power takeoff (PTO) shaft accidents can cause severe and fatal wrapping injuries.

Women are at a higher risk than men of being run over by tractors and other farm machinery, according to data from the US National Safety Council. The increased risk is because women often assist their spouses by hitching equipment to tractors.

Drawbar Connection

The drawbar is the solid metal shaft sticking out the lower rear end of the tractor. It has a center hole near the end to which the U-shaped tongue of the pulled implement is attached with a pin. This connects the tractor to the implement for movement. A single drawbar attachment is referred to as a single-point hitch.

In a single-point drawbar connection, the tractor is backed up to the implement and a pin connects the tongue of the implement to the drawbar of the tractor, much like the connection between a car and a trailer.

- When backing up the tractor to line up with the implement, do so without anyone standing in between the tractor and implement. If your foot slips off the clutch or brake, the tractor could crush the helper.
- Once the tractor is as close as you can manage to the right hitching point, set the brakes and stop the engine so the tractor does not roll away or toward you when you step behind it to insert the connecting pin.
- Once the hitch pin is inserted, secure it with a locking pin or clip so the connection is not unexpectedly lost during work.
- When unhitching any implement, prevent sudden movement by unhitching on level ground or setting sturdy blocks at the implement wheels. Set a large block under the drawbar jack to keep it from sinking into the ground. If the machine sinks, it becomes awkward to reattach and increases the risk of muscle strains.

Three-Point Hitch Connection

A three-point hitch connection employs four components working together: the tractor hydraulic system, the side attaching points, the three lifting arms, and stabilizers. The two lower hitch lifting arms are controlled by the hydraulic system and provide the lifting, lowering, and even tilting of the arms. The upper center arm, called the top link, is movable but is usually not powered by the hydraulic system. Each arm has an attachment point where the implement is connected. The implement has posts that fit through the holes of the attachment arms. The implement is secured by placing a locking pin or clip on the ends of the post. The hitch lifting arms are controlled by the operator using the tractor's hydraulic system, which typically has several settings.

With implements mounted on the three-point hitch, backing in to make the connection is slightly more difficult because you have to line up the two

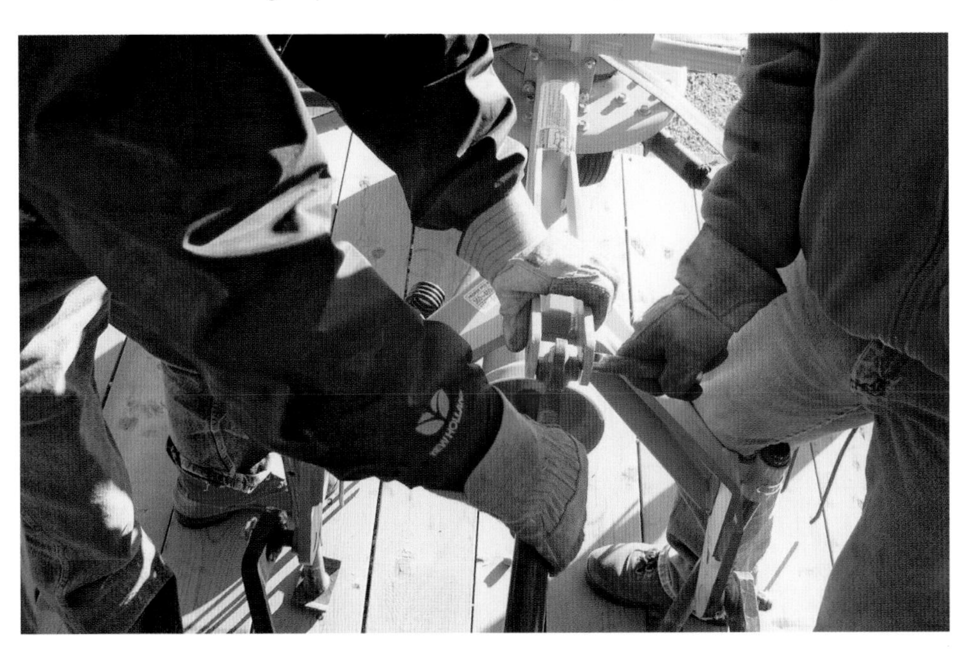

With implements mounted on the three-point hitch, backing in to make the connection is slightly more difficult because you have to line up the two connection points on the lower lift arms. Go slow and make sure no one is behind the wheels while you back up and don't let anyone behind the tractor until you have shifted it into park or have turned the engine off.

connection points on the lower lift arms. Some considerations for using a three-point hitch system include the following:

- When backing up the tractor to line up with the implement, anyone giving directions should stand behind the implement, not between the tractor and implement. If the tractor operator's foot slips off the clutch or brake, the tractor could crush a helper standing between the tractor and the implement.
- Once the tractor is as close as you can manage to the right hitching point, set the brakes and stop the engine so the tractor does not roll away or toward you when you step behind it to insert the connecting pin.
- Once the link pins are inserted, secure them with locking clips so the connection is not unexpectedly lost during work.
- Before moving, test the connection by raising and lowering the implement. Ensure that any implement supports are fully raised to their working positions so a safety hazard is not created by the implement striking the supports.
- When unhitching any implement, prevent sudden movement by unhitching on level ground or setting chocks behind and in front of the implement wheels.

Power Takeoff (PTO)

A PTO is a method for transferring the tractor's engine power to a farm implement to do the work intended, such as chopping hay, spreading manure, or grinding feed. The engine power rotates a splined output shaft called a

PTO shafts that transfer tractor power to implements add another potential area of injury. Risk occurs both from simple pinch injuries when the yoke is being connected, and more serious wrapping injuries if the shaft rotates. Always turn the tractor engine off before connecting the yoke to the spindle.

PTO shaft, to which the farm implement can be easily connected and disconnected. The PTO shaft can rotate at a very high rate of speed to help operate the implement attached. This shaft can also be a dangerous part of the tractor when it is fully operational. Getting clothing, hands, feet, arms, or legs caught in a fast rotating PTO is a recipe for severe injury or death. They can be lethal even at low rates of rotation. They are unforgiving of even a slight mistake on the operator's part. Extreme caution must be used when the PTO is running.

PTO safety shields are sometimes removed for convenience in hitching or break off when the shaft is in use, such as when the shaft is angled beyond normal limits. Always keep shields in place while operating any PTO-driven equipment.

Secure the PTO shield safety chains to solid parts of the tractor. To reduce the risk of wrapping injury, the safety chains keep the shield stationary while the shaft turns inside the shield.

All farm equipment purchased today has PTO protection shields and shaft guards installed at the manufacture points. However, a lot of older farm equipment used today was sold prior to these safeguards, and you will need to look at your equipment or what you are considering to purchase to assess their safety state. To make a safe PTO connection, do the following:

- Make sure the PTO selector is set to neutral. Stopping the engine completely adds another margin of safety. Shut off the PTO whenever moving between areas of work.
- To prevent entanglements, footwear should be fully laced, loose clothing tucked in, and hair tied up or securely tucked under headgear.
- Do not allow anyone (including children or dogs) to remain in the tractor cab or on the operator's station while connecting the PTO unless you stop the engine completely. Accidental engagement of the PTO while a person is near the shaft or implement will be dangerous.
- With everyone clear of the connected shaft, return to the operator's station and engage the PTO at low engine revolutions to test for proper operation.
- **Do not** remove factory-installed PTO shaft guards or shield plates for the sake of convenience. Repair any damaged guards or plates to their previous condition.

Hydraulic Hose Connections

Tractors and other farm equipment use hydraulic fluid that is transmitted throughout the machine by various hoses. The pressure created by the forced fluid through the hoses will turn, lift, lower, and tilt equipment to do their intended tasks. This fluid is controlled directly or automatically by valves through a lever handled by the operator.

Hydraulic hoses are graded by pressure, temperature, and fluid compatibility. Hoses are used where flexibility is needed. Hydraulic hoses generally have steel connectors attached on the ends that are inserted into the corresponding hydraulic outlets on the tractor. Some considerations for hydraulic hose connections include the following:

- Hydraulic fluid under pressure that gets injected under the skin can cause severe, permanent tissue damage even though initial symptoms may be minor. Wear suitable eye and skin protection when working with hydraulic hoses.
- Wipe the hydraulic connector clean before connecting to the tractor. This makes connection easier and prevents contamination from entering the tractor's hydraulic system.
- If the connector will not click into the tractor's hydraulic receptacle, pressure in the line may be interfering with connection. Stop the tractor engine and move hydraulic controls back and forth to relieve all hydraulic pressure within the tractor, and then reattempt the connection.

Hydraulic hose connections can involve a risk of eye and skin irritation or burns or tissue damage from oil spray if a connection is attempted when there is pressure in the lines. Always turn the engine off and then shift the hydraulic engagement levers back and forth to release the pressure in the line before attempting to remove them.

- To prevent safety problems when subsequently disconnecting the equipment, be sure all hydraulic pressure is relieved first. With hoses still connected, lower the implement to the ground or onto supports so no pressure remains in the lines. Shut off the tractor engine and move hydraulic controls back and forth to relieve all pressure. Only then should you disconnect the hoses.

MACHINERY AND IMPLEMENT SAFETY

Older equipment used on farms may be missing the operator's manual and many of the shields, guards, and warning labels it had when manufactured. Equipment purchased at an auction or farm disposal sale may not come with instructions or training for its proper use. It may also have suffered abuse or lack of maintenance, which makes it more dangerous in continued use.

First Step to Improved Safety

The first step to improved safety around machinery and implements is to acquire a relevant operator's manual and refer to it for safe operating procedures. New farm equipment dealers can often access manuals for their older products. Reprints of older manuals can often be ordered or found on Internet auction sites.

Once you have a manual, use the information to get the machine or implement as well adjusted and lubricated as possible before starting work. A well-maintained machine or implement is less likely to break down or plug up, and fixing or unplugging machinery is often when accidents happen.

Engine Power

When adjusting or repairing a machine, all power should be completely shut off by stopping the tractor engine or on-board engine. There is much less risk of accidental movement of the machine and subsequent injury of the operator.

If it's impractical to completely shut down the engine, be aware of the possible risk of accidental engagement and do what you can to protect yourself. Consider ways you could reduce exposure to potential danger points, such as clearing baler blockages with a pitchfork instead of your hands or inserting a hefty wooden block in the drive mechanism to prevent accidental movement while you are near the machine.

The Frustration Factor

Many farm machinery injuries are related to blockages and breakdowns that occur during use, such as clearing hay from an improperly functioning baler. When these blockages and breakdowns occur, the operator is likely to be in a frustrated or agitated frame of mind, especially if the problem is occurring repeatedly or has not been resolved by normal measures.

When frustration builds up, reasoned judgment often goes out the window. Alternatives that are much less risky but might take a little longer are

GENERAL SAFETY PROCEDURES

Always, always, always follow these top ten safety procedures when using machinery on your farm. This is *not* a complete list of all the safety procedures you need, but it's a good start.

1. Wear suitable equipment such as heavy gloves, eye and hearing protection, and boots with nonslip soles.
2. Observe warning labels.
3. Before attempting to unplug, clean out, adjust, repair, or lubricate any unit, turn off the source of power and physically block any parts that may suddenly move when an obstruction is removed.
4. Once power is shut off, wait for the mechanism to come to a complete stop before approaching the machine.
5. Do not allow children or pets to remain in the tractor cab or operator's station. They could accidentally engage the machine drive or PTO.
6. Secure any dangling clothes, hair, or jewelry that could get tangled in the machine.
7. Never attempt to engage or disengage a v-belt drive by pulling on the belt with your hands.
8. If the machine is held up by hydraulic power, set blocks to prevent unexpected movement before you crawl underneath the machine.
9. Before you move a machine or engage the drive, be sure you know where any bystanders, especially children, are located.
10. Avoid moving in reverse unless you can be certain there is no one behind you.

overlooked or quickly rejected, even though the delay they bring is much less than the delay entailed by an injury.

When you are frustrated, your lack of focus can more easily lead to a work-stopping or life-ending accident. In times of frustration, a personal checklist of countermeasures can be an alternative that improves both short- and long-term personal safety.

Machine and Implement Safety Highlights

The following alphabetical listing of farm machines and implements reviews some of the key dangers and safe procedures to follow in use. When you purchase a new or new-to-you machine, review the operator's manual for both your safety and your family's safety.

Auger

A rotating screw auger carries feed, grain, or other material up a tube. It also creates a crushing or cutting hazard if hands or feet get caught between the screw and the tube. Auger intakes should have grates with openings only wide enough for the material, not wide enough for hands and feet.

- Do not reach in any moving auger.
- In windy conditions, be careful about raising the auger to full height. Strong winds can tip it over.
- If the auger is raised and lowered with a cable winch, do not let the crank handle spin freely and do not try to stop it with your hand if it does get away.
- Contact with an overhead power line can cause an electrocution injury. Check overhead whenever lifting or transporting the auger.

The rapidly turning screw, or flighting, can quickly sever body parts. Maintain a safe distance when working with grains being fed into an auger like this.

Baler (Rectangular)

- Never feed material into the baler by hand, such as rebaling a broken bale. Spread the material out on the ground and let the baler pickup gather it.
- Do not try to pull twine out of the knotter while the baler is running.
- After the PTO is off, wait until the flywheel has completely stopped turning before approaching the machine.
- Gathering forks, plungers, and knotters in a conventional square baler may remain under tension if the machine is accidentally plugged during use. When clearing the machine, use blocks in the mechanisms so that the release of tension does not cause unexpected movement that could entangle you in the machine.

If you need to work on a round baler, always engage the safety lock or close the hydraulic valve before moving underneath the rear bale ejection door.

- Use only the correct type of shear bolts in the baler. If shear bolts fail repeatedly, fix the cause and not just the symptom. Refer to the manual for the location of points protected by shear bolts.

Baler (Round)

- Never feed material into the baler by hand, such as rebaling a broken bale. Spread the material out on the ground and let the baler pickup gather it.
- Use only the correct type of shear bolts in the baler.

If you need to clear plugged material from the baler pickup, stop the PTO power and block the gathering forks so they cannot accidentally move when the blockage is removed. Remove the block before restarting.

Chisel Plow

- Engage mechanical safety locks or use strong blocks whenever it's necessary to climb under the machine.
- If you are welding cracks on frame members of the chisel plow that are made of square or round tubing, hot gases can build up inside the tube and cause an explosion. Before starting the weld, make sure there is a vent for gases to escape freely.
- When transporting the cultivator with its wings folded up, watch out for overhead power lines.

In difficult harvest conditions such as tough straw, heavy weeds, or uneven crops, the infeed area of a combine can easily become blocked and require removing the material by hand. Be certain that both header and threshing drives are fully disengaged before entering the table area.

Corn Picker

- Older model corn pickers can frequently plug if ground speed is too fast or slow. Stop the picker mechanism and turn off the tractor before unplugging material.
- Snapping rollers travel at about 12 feet per second, which means a farmer holding onto a stalk 2 or 3 feet away has less than a half-second to react. Always turn off the picker and tractor before dismounting to unplug it.

Cultivator

- Engage mechanical safety locks or blocks whenever it is necessary to climb under the machine.
- When transporting the cultivator with its wings folded up, watch out for overhead power lines.

Disc

- Disc blades have sharp edges, so be careful when working and making adjustments around them.

Pickers are frequently plugged if ground speed is too fast or slow. There is a major entanglement risk if you attempt to clear out plugged material without stopping the picker and shutting off the tractor or picker drive. Never attempt to dislodge plugged materials while the picker drive is running. These were often referred to as "widow-makers" for obvious reasons.

Packer-type drills are less subject to tipping because the weight is born by dolly wheels at the front and the packers at the rear. Periodically inspect the catwalks located near the rear. When you fill seed boxes, your attention will be focused on the job, which makes it easy to slip from a catwalk in poor repair.

When removing a gang of discs to change pans or bearings, be aware that the gang can be very heavy. Lower the gang using jacks or ropes attached to the frame.

- Wear gloves when changing blades. Lacerations may occur when pushing on the heads of the plow bolts that hold the tips to the shanks. If you are using an air wrench, the head can spin suddenly and slice your finger.
- When transporting the disc with its wings folded up, watch out for overhead power lines.
- If the disc is mounted on a three-point hitch, turning the tractor swings the disc the other way, so turn carefully to avoid hitting objects.

Drill (Seed)

- Lower the grain drill to the ground before servicing or adjusting the drill, such as when removing bags from the openers during calibration or changing soil openers.
- No riders are allowed while the grain drill is being operated. The catwalks at the rear of the drill are for use only while the drill is stopped. A sudden jolt or bump while moving could cause the rider to fall off.
- When filling the drills, be sure the catwalks are clear of dust, grease, and obstructions to reduce the risk of slips or falls.
- When transporting the drill, be sure any transporting device (end wheels) are properly positioned and secured with locking pins.

Safety hazards in a portable grinder-mixer mill involve crushing by the hammer mill section, cuts by the mixing auger, or entanglement in the slinger or PTO drive. Breathing dust when cleaning the bin is also a hazard.

Feed Mixer-Mill

- Make sure the PTO shaft has all guards and shields installed, and keep clear of the shaft in operation.
- Avoid overloading the grinder, which can result in plugging that has to be cleared by an operator.
- Change mill screens or clear plugs only after the hammers have come to a complete stop.
- Slippery, uneven ground or horseplay around the intake hopper may cause the operator to make hazardous contact with the machine.

Forage Harvester

- Block the header securely before climbing underneath.
- Know and follow the manufacturer's procedures for sharpening the knives. Stand in the correct position and do not lean over the sharpening mechanism.
- When sharpening, wear the recommended personal protective clothing, including eye and ear protection.
- Before opening any guards or shields, allow the cutting mechanism and drives to come to a complete stop.
- Know and follow the correct procedures for clearing a chute blockage.
- Use the reversing drive to clear blockages. Before you attempt to clear any blockage by hand, make sure the cutter-head is stopped, the PTO is disengaged, and no one is near the tractor controls.
- Follow maintenance and lubrications schedules. A poorly maintained machine is more likely to perform badly and cause the operator to make adjustments.

Forage harvesters used on small farms are often older pull-type models rather than the large self-propelled units that are similar to combines. Typical hazards on the pull-type machine include entanglement with the PTO or moving drive mechanisms, contact with the exposed rotating cutter-head while sharpening, getting trapped or injured by a falling header, or falling while adjusting the discharge spout.

Forage Wagon

- Before approaching the machine for inspection, adjustment, or repair, disengage the PTO and allow the unloading beaters to stop completely.

- If it is necessary to climb inside the machine, be cautious about slippery areas that could cause falls that lead to entanglement in the conveyor and/or unloading beaters.
- When disconnecting the forage wagon, block the wheels.

Manure Spreader

- PTO shafts and the side shaft that transmits power to the beaters on conventional spreaders are key hazard areas. Keep them shielded and re-install shields after repairs.
- Always shut off power to the spreader before scraping the sides or entering the spreader box.
- To help prevent entanglement in shafts, chains, beaters, or augers, secure any loose clothing or hair.

Mower

- Before working anywhere near the conditioners at the rear of a mower-conditioner, disengage the PTO and let the mechanism stop completely.
- Keep bystanders away from a working three-point hitch rear-mounted mower, as it can propel stones and debris out the back and sides at high speeds.
- If working under a mower, use blocks to prevent the mower from falling on you. Hydraulic lifts alone are not sufficient for holding up the machine while you are under it.

These machines are only a few that may choose to use on your farm. Although they may have different operation mechanisms, always remember the basic rules for safety.

DUST, MOLD, GASES, AND CHEMICALS

Country living is often associated with wide open spaces, fresh surroundings, and clean air. It is true for the most part, but awareness of hazardous and polluted periods and occasions will help keep you and your family alive and healthy. Hazards can involve dust, molds, gases, and chemicals and these can be serious threats to any farmer's health.

Dust and Molds

In many cases the effects of dust and molds can be mitigated by the simple use of an air barrier such as a dust mask that is placed over your nose and mouth. Masks can be form-fitted to the shape of your nose and mouth to prevent a high percentage of dust from entering your nasal cavity and lungs. They range from the inexpensive kind sold in many farm stores to the more expensive respirators.

Farmer's lung, or hypersensitivity pheumonitis, is an incurable, allergic lung disease caused by the inhalation of spores found in moldy crops,

Dust and mold spores are encountered in many agricultural activities and are often associated with respiratory illnesses.

such as hay, straw, corn, silage, grain, and tobacco, or in bird-breeding and mushroom-growing businesses. A milder response from exposure to relatively low levels of dust is marked by coughing, shortness of breath, sweating, sore throat, headache, and nausea or fatigue. A chronic response can develop after persistent acute attacks and recurring milder responses.

Organic toxic dust syndrome is a flulike illness due to the inhalation of grain dust with symptoms including fever, chest tightness, coughing, and muscle aches. The inhalation of the grain dust may occur in an agricultural setting or from covering a floor with straw.

Gases

A variety of potentially toxic gases are produced during many routine agricultural operations. When silage is put into a silo, plastic wrappers, or covered silage bunks, natural chemical processes that ferment the materials and preserve them also emit gaseous byproducts including nitrogen dioxide. This is a yellow, brown, reddish, or orange cloud that may have a bleach-like odor when it is in confined areas and in high concentrations. When this gas comes in contact with moisture in your lungs, it can form nitric acid strong enough to severely burn the insides of your lungs and result in severe injury or death. Keep children away from silage storage areas, and you should stay

Motor fuels and lubricants present a potential soil and water contamination hazard if they aren't stored and transferred with the proper procedures.

out of any silo that has recently been filled.

Many silos have storage rooms attached at the bottom. Make sure that the doors are open for several days after to provide fresh air access into the room, and avoid entering the room yourself. Gas pockets can linger for several days before finally dispersing. The same holds true at the top of the filled silo.

Gas is heavier than air and will settle in low spots. If you enter a silo too soon after filling, you may step into a pocket of gas that has settled in a low area of the silage. Always blow fresh air into the silo with your blower for at least twenty to thirty minutes before entering and leave it running while you are in it. Remember that gas can linger in the silo chute and you may encounter it while you climb up the inside. Before entering a silo for the first time after filling it, let people know where you are going or have someone below with whom you can talk while you are working there. You need to be extremely careful because help is often too long and too far away to reach you in case of emergency, and it is difficult to remove a body from a tall silo to give medical assistance.

A manure pile will produce gases as the manure rots and composts. Typically these do not cause many problems unless the pile is contained within an enclosure. Any enclosure will need to be ventilated before entering to remove the manure.

A greater risk is a liquid manure pit. Gases get trapped in small bubbles in liquid manure and are released when the manure is agitated or pumped. Many deaths have been caused by farmers entering what they thought was a safe area because the pit was an open air structure.

Manure gases contain hydrogen sulfide, which is a clear, colorless gas. It is heavier than air and tends to pool in low areas. Although it may sometimes smell like rotten eggs, a high concentration of this gas can temporarily paralyze the nerves in your nose so that you are unable to smell it. High concentrations can cause immediate stoppage of breathing through paralysis of the diaphragm. Never enter a manure pit by yourself, and all who need to enter should wear an oxygen mask and supply.

Carbon dioxide ammonia and methane are other gases produced by manure bacteria. Each is equally dangerous, and you need to use caution when working in areas containing liquid manure.

Carbon monoxide is a clear, odorless, and very deadly gas formed during combustion in equipment, such as internal-combustion engines and space heaters. Be sure the ventilation in your shop or shed is sufficient for air replacement cycles to avoid problems.

Paints and solvents can also give off various hazardous and/or flammable gases as they dry, so be sure to work with proper protective equipment and plenty of ventilation when using these materials.

Chemicals

Many farms use a variety of chemicals even if no pesticides are purchased or used. Chemicals used on farms may include motor fuels, lubricants, liquid sprays, powders, and a host of other forms. Proper use, storage, and disposal of chemicals protect you and others from accidental exposure. Proper storage and disposal are particularly important for protecting inquisitive children exploring the farm. Proper handling is also vital to preventing injury to farm animals, water, and the environment.

Always read the labels for use and disposal information. Whenever using any chemical substance wear correct personal protective gear, such as appropriate respirators, eye protection, impermeable aprons, and gloves.

If the chemical is purchased new, it generally comes with a copy of the Material Safety Data Sheet (MSDS). The MSDS outlines any health hazards and appropriate precautions to take, along with other important information like handling and disposal procedures.

DANGER SIGNS IN LIVESTOCK

- Raised or flattened ears
- Raised or rapidly lashing tail
- Raised hair on the back
- Bared teeth
- Rolling eyes
- Making unusual or distressed cries
- Climbing onto the backs of other animals to escape
- Snorting, tossing the head, and pawing the ground
- Stiff-legged gait or posture
- Any history of previous aggression

LIVESTOCK

Raising farm animals such as cattle, horses, sheep, pigs, and goats is the dream of many who are attracted to country living. Over time, these animals often seem like a pet or part of the family, especially to children and visitors. But all livestock should be considered unpredictable. They have important instinctive behavior patterns such as rejoining the herd, maintaining herd hierarchy, maternal protection of newborns, male attack of intruders, and reaction to sudden movement and noises beyond the range of human beings.

Consider the weight of the animal and what that could mean if all that mass collides with you. A 1,200-pound horse or 1,500-pound bull pressing you against a fence is bad enough, but if that animal is also moving at the time, that is a lot of kinetic energy to get absorbed by fragile human bone and tissue. Even a 200-pound pig, moving at top speed and accelerating amazingly fast, can knock you off your feet when you're trying to catch one.

Keys to keeping you and your family safe around livestock include having knowledge about instinctive reactions of different animals and how to use that knowledge to advantage when herding, feeding, treating, or handling livestock. You also need to know how to keep livestock areas free of hazards, such as improperly designed chutes and lanes, lack of human escape routes, and sharp protrusions.

Livestock with Young

Mothers of newborn animals can exhibit a very strong maternal instinct that includes being very defensive of their young, and therefore they can be difficult to handle if they must be moved from place to place. When possible, let the young stay close to the adult during handling.

Dealing with Animal Danger Situations

Learn the signals that animals use when they feel threatened and are ready to attack. The sidebar gives a list of danger signs to learn. Though this isn't a complete list, it covers many of the instinctive behaviors related to safety. Most animals have a strong territorial instinct and develop a very distinctive, comfortable attachment to familiar areas, such as pastures and buildings, water troughs, worn paths, and feed bunks. Forcible removal from these areas may cause animals to react unexpectedly.

Large animals can generally see at wide angles around them, but there is a blind spot directly behind their hindquarters. Any movement in this blind spot will make the animal uneasy and nervous. The sense of smell is extremely important to animals, especially between females and newborns. Handling facilities should be only one color since all species of livestock are likely to balk at a sudden change in color or texture.

Cows, Bison, Pigs, and Sheep

- These are herd animals and are calmed by being in groups. Separation from the group can lead to a panicky flight behavior. Conversely, the

A mother's maternal instinct is generally beneficial, but it can also create an extreme safety hazard. An innocent situation can turn dangerous if she perceives a danger and attempts to protect her young. Keep a watchful eye on any new mother of any species.

Be alert around any horse until you determine its attitude toward you or any family member. Newly purchased horses can become agitated as they learn to understand their new surroundings.

herd can be moved by concentrating on the dominant animal so the others will follow.

- Their instinctive tendency is to move from a dimly lit area to a more brightly lit area, provided the light does not hit them directly in the eyes. A spotlight directed on the ramp will often help keep the animals moving.
- Always announce your presence when approaching. For cows, lightly touch the animal rather than making the first contact with a bump, shove, or a poke with a stick.
- Herd pigs with a lightweight panel to prevent them from trying to make a sudden dash past you. Quietly and gently make yourself known to avoid startling the pig. A knock on the door or rattling the door handle may signal food rather than danger.
- Sheep have difficulty picking out small details, such as the open space created by a partially opened gate.
- Always keep a close eye on rams (male sheep) that often butt if your back is turned.
- A sheep can be immobilized for safe handling by sitting it upright on its rump.
- Small pigs can be handled safely by grabbing the back legs and raising the animal off the ground.
- Cows and pigs have poor depth perception and difficulty in judging distances. Cows may not be able to distinguish between a real cattle guard and parallel striped painted on a road.

Horses

- Ears point toward where attention is focused. Ears that are flattened backward warn you that the horse is getting ready to kick or bite.
- Always work with calm and deliberate movements around horses. Nervous handlers can make horses nervous, which creates unsafe situations.
- Be careful when approaching a horse that is preoccupied, such as when its head is in a hay manger. Speak to the horse to get its attention and wait until it turns and faces you before entering the stall.
- Always use a lead line to prevent getting a hand caught in the halter if the horse unexpectedly moves.
- To lead a horse through a doorway, step through first, then quickly step to the side and out of the horse's way.
- Never wrap any piece of tack around your hand because it could tighten and injure the hand if there is sudden movement.
- After you remove the halter, make the horse stand quietly for several seconds before completely letting it go. This pause will help prevent the horse from developing a habit of bolting away and kicking at you.
- Do not climb over the lead line of a tied horse. The horse may pull back and jerk the line tight, causing you to fall.

FARM FIRE SAFETY

Farms are typically far away from emergency services, such as firefighters. Response times may vary but the farther away from a rural volunteer or professional firefighting service the more vigilant you need to be in minimizing fire threats.

Fire Readiness

The first step in fire readiness is having your access road and farm number clearly marked. This is typically done by the town or county government, but having clearly marked signs will help first responders to find your farm. Other steps include the following:

- Have simple discussions so everyone knows what to do or where to go in case of a fire. It is especially important to protect children, so you and firefighters don't spend critical moments searching for them.
- Human safety, including your own life, must be your first priority. Make sure you, your family members, and employees are safe.
- Call the fire department immediately and let the experts take control. People have been seriously injured or killed when trying to save animals, grain, or equipment, forgetting that smoke and toxic fumes can kill them in seconds.
- Have a list of what flammable items are stored where so firefighters know what to be prepared for and what to protect themselves against. Location of fuels, lubricants, and pesticides is especially important.

Farm Access

- Be sure everyone in your family who is capable of calling the fire dispatcher can give clear, concise farm number information. Keep these by your main phone.
- Keep your yard as clear of machinery as feasible so in the event of fire there will be room for fire trucks to turn and set up, and you won't need to spend valuable time moving equipment out of the way.
- Have information available about any reliable, accessible water source such as a local stream or pond. It is not practical for rural firefighters to carry large volumes of water to fires.

Livestock Behavior during Fires

- Animals may panic and refuse to leave a burning building, or in some cases they may run back in after being let out. Don't become trapped by repeated attempts to evacuate them.
- If you are able to evacuate animals, be sure you are not leading them toward a dead end, such as a gate that won't open outward.
- Immediately call a veterinarian to examine any animals suffering from heat, smoke inhalation, or burns.

Many fires are accidentally started when the wind stirs up a burning barrel. Keep a screen atop the barrel if you leave it unattended.

Hay (loose or bales) that is stored at too high a level of moisture can start to heat up and eventually burn. Make certain that all hay is properly dried before you put it in a building or outdoor stack. Cover outdoor stacks to prevent infiltration of rain that can lead to mold and subsequent heating.

Hay and Grain Storage

- If hay is slowly smoldering in an upper level of a barn, call the fire department and begin evacuation if possible. Do not try to throw smoldering hay out a window or door because exposure to oxygen will make the fire flare up.
- If you see or smell smoke coming from the hay, place boards or plywood on the hay before walking on top of it to probe for hot spots. The boards will prevent falls into burned-out cavities below the surface. A rope as a lifeline is recommended.
- Silo fires are extremely dangerous and you should stay out of any silo chutes as the toxic smoke will linger. Let trained firefighters handle tower fires.

Construction Planning

- When constructing any major farm building, the site plan should allow for adequate spacing between buildings to prevent the spread of a possible fire.
- Locate new buildings at least 40 feet away from above-ground fuel storage tanks to minimize the potential for spread of fire. Consider a longer distance if the buildings are in line with prevailing winds.
- When constructing a new building, include fire-prevention features such as fire doors, a firewall between hay/bedding storage and stabling or work areas, flame-retardant or fire-resistant materials, fire-retardant latex paint, smoke detectors, fire alarms, and automatic heat-sensitive sprinkler systems.

Electrical Safety

- Fixtures for fluorescent lights should have dust- and moisture-resistant covers. Incandescent bulbs should have globes sealed against moisture and dust plus a metal cage to prevent accidental breakage. Older barns with light bulbs hanging from a wire are a fire waiting to happen.
- Make sure the wiring and fuses/breakers can meet your current needs without overloading the electrical system. Overloads cause excess heating, which can lead to fire in the building.

- Use accredited electricians for any rewiring projects.
- Electric motors in livestock facilities should have moisture/dust-proof on/off switches. Motors should not be within 18 inches of any combustible material such as hay or bedding.
- All major or tall buildings should be equipped with lightning rods or other suitable lightning grounding provisions.

Fire Extinguishers

- Keep a fire extinguisher in every building where it's practical so that if a fire develops you have one at hand.
- Read and understand the directions of any fire extinguisher you buy so you know how to use them correctly when needed.
- Service all extinguishers at recommended intervals.
- Always think "safety first." Know your limits. Fire extinguishers cannot do the job of a local fire department. If a fire gets beyond your control while you are waiting for help, get out of the building and account for all family members.

Farm Fuel and Lubricants

- Be sure all containers for flammable and combustible liquids are clearly and correctly marked so that they do not end up near sources of heat or sparks.
- Never store fuel in breakable containers.
- Immediately take care of leaks or deterioration in fuel storage units.
- Let only qualified personnel cut, weld, or solder a fuel tank. It is better to replace tank if you are unsure about its stability.
- Do not keep gasoline inside your home or transport it in the trunk of an automobile or RV. If gasoline must be transported, use only an approved container and roll down windows so moving air can sweep away vapors.
- Diesel, motor oil, and grease require more heat than gasoline to ignite, but once ignited they will burn long and hot. Keep them away from heat and sparks. Remember that gasoline and aerosol propellants will explode and ignite more quickly than diesel fuel.

Portable Heaters and Heat Lamps

- Portable heaters should not be used in the barn area. Allow for adequate fresh air flow to increase oxygen levels in the area where heaters are used.
- Make sure heaters have a shutoff device that activates if the unit is knocked over. Place the unit where livestock cannot knock it over.
- Heat tapes and water tank heaters must have a thermostat and be adequately protected so that horses, livestock, cats, dogs, or rodents cannot chew them. Chewing can result in electrocution, shocks, and/or fire.
- Heat tapes should be protected with a fire-retardant insulation material.

General Fire Safety

- Smoking should never be permitted in any building where combustible materials are located or stored.
- Exit doors should be clearly marked.
- Aisles should be raked or swept clean of hay and bedding. Vacuum cobwebs and dust regularly. Wipe dust/dirt off light fixtures, outlet covers, switches, and panel boxes.
- Be very careful about burning trash or paper, especially in windy or dry conditions. Obtain applicable burning permits for large burns.

CHILDREN'S SAFETY

Children growing up on farms are often curious about their surroundings and will explore areas that may prove hazardous for them. Although you can never completely childproof a farm, with attention to some of the prime accident areas, you can eliminate a lot of needless risk. In many cases, the changes are beneficial even if children aren't around. Key areas and precautions areas include the following.

Canals, ponds, streams, lagoons, or other bodies of water are endlessly fascinating for children. But steep banks, strong currents, or deep mud in farm water bodies can turn a play area into a death trap. Always supervise children, especially near water.

Water Bodies

- Supervise children at all times around water, just as you would at the pool or beach.
- Wherever practical, fence in all ponds and lagoons with self-closing, self-latching gates as part of the fence. Post "no trespassing" signs and explain to children what the sign means. Make rescue tools available near water such as a rope, flotation rings, a long pole and/or a loud whistle to summon help. Show all children how to use these tools.
- If you live in an area where water bodies freeze over, explain to children, at their level of understanding, basic ice hazards and ice rescue. Your library, local Red Cross chapter, and fire departments can provide information.
- Irrigation canals should be considered off-limits to both children and adults, especially in main canals. Steep banks and currents make it difficult to climb out of the water.

- Warn children to stay away from wells, spillways, gates, intakes, motors, and other system parts.
- Cover the tops of all wells or septic systems with sturdy materials and do not allow children to play near them.

Machinery

Farm machines can be fascinating to children, but small children have no place around them without close supervision and secure seating. Fenders or laps are not good enough. Youths learning to operate machinery also need a secure trainee platform and careful training even in activities you think might be taken for granted, such as operating tractors at road speeds.

Practical Safety Precautions around Farm Machinery

A few simple, practical precautions should be part of your day when children are around.

- Get a definite, visual confirmation of where children are before moving a tractor, implement, or other piece of equipment. Unless you can actually see that they are in a safe position, small children may be out of sight behind wheels, tracks, and other machine parts.
- Never allow riders on a tractor, implement, harvester, ATV, or other piece of equipment unless a secure operator's seat or platform is provided. A child who falls off due to a sudden jolt or turn is at extreme risk of being run over or entangled. Due to the size of the machine, the operator may not even see the accident in time to prevent it.

Many serious falls happen because children are fascinated by climbing up onto high places such as sheds, haystacks, machinery, and fuel tanks.

- Children riding on fenders or laps may interfere with the operator's access to machine controls and increase the risk of an accident.
- Never allow riders in the back of a pickup truck without seats and grab rails. Sudden vehicle movements can shift riders.
- When dismounting from a tractor make sure all riders dismount and remain at a safe distance if making adjustments or repairs to an implement. Unexpected accidental engagement of the implement drive by a rider may cause injury or death to the operator.
- Take the time to explain what you are doing so children do not have to satisfy their curiosity about the job by getting too close to hazardous machinery parts.

Falls—Preventing Farm Falls for Children

- When not using a ladder, lay it down on the ground or secure it on a wall storage rack. Besides being a safety benefit, this prevents the ladder from getting knocked down and damaged by wind or animals.
- On machines such as harvesters with elevated access platforms, keep platforms clear of mud, tools, twine, or other obstructions that could lead to slips and falls.
- Keep the area around haystacks clear of protruding stakes or other objects that could impale a falling person.

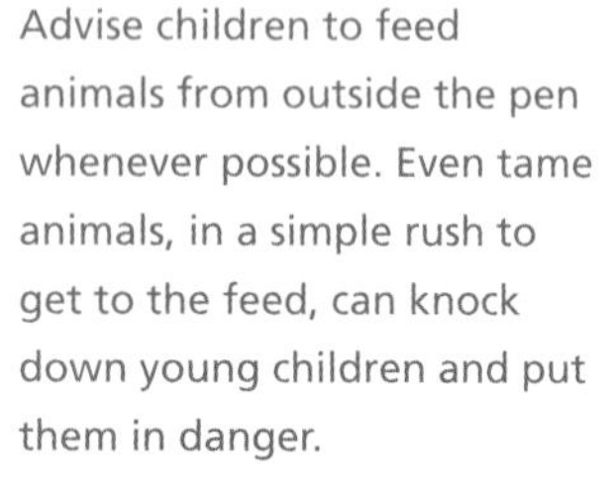

Advise children to feed animals from outside the pen whenever possible. Even tame animals, in a simple rush to get to the feed, can knock down young children and put them in danger.

Farm Animals

Most children are fascinated by animals, and they may assume farm livestock are pets to play with. Baby animals are even more irresistible. However, all farm animals should be considered unpredictable and must be treated with respect and caution. The large mass of a farm animal, such as a cow or horse, means accidental or aggressive contact with a small child will result in extensive injury. Most children do not have sufficient strength or speed to quickly scramble out of the animal's pen in a dangerous situation. Some precautions for children near farm animals include the following:

Children may view farm animals as pets, so it's important to teach them to treat animals with caution.

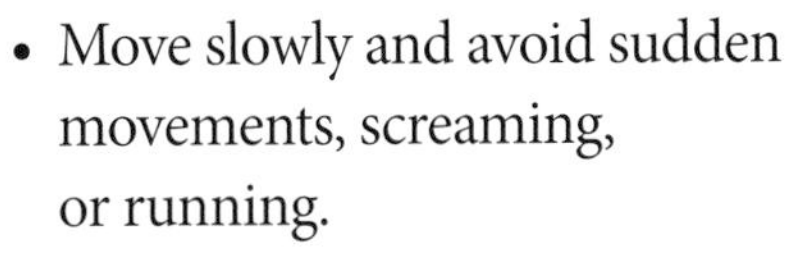

- Move slowly and avoid sudden movements, screaming, or running.
- Wherever possible, approach animals in the company of someone already familiar to the animal. Seeing someone they recognize reduces stress in the animal.
- Always approach animals from the front so they can see what you are doing.
- A mother animal is usually very protective of her young and may vigorously defend them against anyone she sees coming close.
- Avoid contact with bulls, boars, rams, and other male animals. They tend to be naturally aggressive.

Where to Get Information

Your county agricultural extension office can provide a wealth of practical information on farm safety. Make use of their services.

Chapter 6
Harvest, Preserve, and Butcher

Each season on the farm has its share of positive characteristics, but perhaps the most heartwarming and satisfying season is that of harvest time. The long months of planning and the hours upon hours of work finally culminate in that celebratory period in which the bounty of the summer is harvested and preserved.

In this chapter, we'll explore a variety of ways to savor the harvest and preserve it for the months to come.

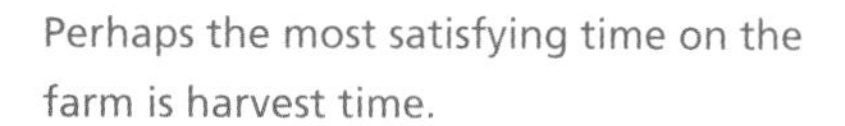

Perhaps the most satisfying time on the farm is harvest time.

HARVESTING AND GRINDING GRAIN

Harvesting grain doesn't necessarily mean that you need to run out and invest in the purchase or rental of a combine, thresher, and other pieces of expensive harvesting equipment. If you're growing grain on a small scale, it's entirely feasible to harvest the grain using hand tools. Harvesting grain by hand is an amazing way to get connected with your land and the traditional methods of farming. Machines such as combines are undeniable time- and labor-savers, but the initial cost of such a substantial piece of equipment is likely too high for the average small-scale farmer.

If you're aiming to turn a profit with your grain-raising endeavors, you'll likely need to harvest a minimum of three to five acres of any particular crop, and it's simply not feasible to harvest this volume of grain by hand. But for home use, small spaces and hand tools are the perfect combination!

For commonly grown grains, such as wheat, rye, or barley, you can confidently and comfortably harvest on a small scale using hand tools. Grains that are more unusual—such as quinoa or millet—are a bit more complicated due to their small size, and some grains, such as spelt or oats, must go through a hulling process, which adds another step to harvesting. Hulling these grains can be easily accomplished with a machine called a huller, but it is also possible to manually hull these grains.

Before you begin harvesting, be sure that your grain is at the proper stage of maturity—you don't want to harvest grain too early or too late. For

If you've got just a small swath of grain that you've grown for home use, it makes more sense to harvest it by hand using a sickle or scythe than to rent expensive equipment like a combine and thresher.

After you cut the stalks, gather them into bunches and tie them to form sheaves, and lean the sheaves together in small groups to form shocks. Dry them for one to two weeks.

example, wheat—like many types of grain—goes through three stages on its way to maturity: the **milk stage**, in which liquid is still found in the kernels; the **dough stage**, in which the kernels grow firmer but are still able to be dented; and the **mature stage**, in which the kernels are hard and have turned golden. Harvesting of wheat usually takes place just before the grain reaches the mature stage—sometimes known as the **late dough stage**.

The Harvesting Process

Harvest your small-scale grain using a sickle, a scythe, or blade trimmers, or (if you're harvesting a very small amount) you can even use shears. Watch the weather forecast and wait for a promising period of dry, sunny weather before you begin harvesting.

After cutting the grain, gather the cut stalks into piles and tie them together to make **sheaves**. Dry the sheaves in groups—known as **shocks**—with the heads pointing up. This can be accomplished indoors or out, but if you have the space to bring the sheaves in, it's better to keep them out of inclement weather.

After the sheaves have thoroughly dried—approximately seven to fourteen days; you can tell the grain is dry when the kernels become crunchy—it's time to begin threshing. **Threshing** is the process of removing the wheat kernels from the plant, and you can accomplish this in a number of ways. Traditionally, a tool known as a **flail** was used in the threshing process, but nowadays, some small-scale farmers use makeshift equipment to achieve similar results. In the baseball bat method, you place the sheaves on a sheet on a hard surface and then lightly bang the heads of the grain with a plastic baseball bat, effectively loosening the grain. It's also possible to use a small length of garden hose. Your ultimate goal is to separate the kernels of grain from the heads of the plant in preparation for the **winnowing** process.

When threshing is complete, you'll be left with a pile of grain that is mixed with an assorted conglomeration of chaff and straw. The next step is a process known as winnowing, which is the separation of the chaff and bits from the grain. Picking out each piece of chaff by hand is a tedious and time-consuming task, so you'll want to speed things up by winnowing.

Winnowing is a sweetly simple process. Take two buckets and slowly pour the grain and chaff from one bucket into the other. If you do this on a lightly breezy day, the wind will carry away the bits of chaff as you pour the grain from the first bucket into the second one. If you must complete your winnowing tasks on a day without a breeze, you can use a small electric fan to aid the process. You'll want to winnow your grain as many times as necessary to remove all of the chaff.

Grinding Grain

Once you have successfully harvested your grain and it's thoroughly dry and ready to use, you'll want to think about grinding it to make it into flour. For grinding, you'll need either an electric grinder or a manual grinder (also known as a grain mill). It's possible to use a blender or to purchase a special attachment for an electric stand mixer, but an electric or manual grinder is generally a more effective choice.

An electric grinder quickly and effortlessly grinds your grain into flour in varying textures. A hand-cranked grinder is obviously more labor intensive and requires a bit more muscle and effort, but it is entirely manual and doesn't require electricity. Depending upon your desire to pursue a self-sustainable lifestyle, you may or may not be interested in pursuing the hand-cranked option.

Two additional factors may influence your choice: price and speed. Electric grinders can be somewhat more expensive than manual, but the prices vary on both versions depending on the style and quality. Top-of-the-line grinders—electric and manual—command a higher price, so it might be wise to evaluate your long-term interest in grinding your own grain before investing in high-end equipment.

As for speed, an electric grinder can obviously make short work of turning grain kernels into flour, while a manual grinder is a decidedly slower process. Bear in mind that the manual grinder will work even when the power goes out, which is a nice benefit. From a strictly aesthetic viewpoint, there's something rather endearing and pioneerish about grinding your own flour using a hand-cranked grain mill, and the added time for the grinding process can be seen as an opportune period to catch up on some reading or family conversation.

Regardless of the model that you choose, you'll want to be sure that it is capable of producing the type of flour that will suit your needs. Some grinders are unable to produce flour that is super fine, so if that's your aim, be sure to find a grinder that is able to do the job.

PRESERVING FRUITS AND VEGETABLES

Is there any sight more thrilling to the farm enthusiast than a colorful row of canning jars, filled with the painstakingly preserved produce of your garden's bounty? There's a wondrous feeling that comes from the knowledge that you can continue to benefit from the fruit of your labors long after the summer season has passed.

Canning is not an easy process, but it's a good way to preserve your farm's harvest for the coming months.

But as rewarding as it is to preserve food for future use, it's not necessarily an easy process. The necessary steps to prepare food for canning, drying, pickling, or freezing can be time-consuming and sometimes complicated—it's both an art and a science. Many people find great enjoyment and satisfaction in canning and preserving for this very reason, and it's a wonderful way to put your garden produce to good use during the winter months.

A note on preparation: regardless of the type of preservation that you're using, it's always wise to select the very best fruits and vegetables. Choose examples that were picked at the peak of perfection, are not squishy or overripe, and do not have blemishes or other defects. This will put you on the right track to ensuring the quality of your preserved products.

Canning

During the canning process, jars of food are heated to high temperatures in order to kill bacteria, mold, and toxins, thus sterilizing the food so that it can safely remain in unrefrigerated storage for several months or up to a year.

Canning uses two types of processing: one process for high-acid foods and one process for low-acid foods. High-acid foods—like fruits, pickles, and tomatoes (with added acid)—only require being heated to a

High-acid foods like fruits and tomatoes can be canned in a hot water bath canner, as they only need to be heated to 212°F.

All vegetables and all meats are low acid foods and must be heated to 240°F and canned in a pressure canner.

temperature of 212°F, and this can be accomplished with what's known as a **hot water-bath canner**, which utilizes a large pot with a lid and a rack.

Low-acid foods—which include all vegetables and all types of meat—must be heated to a significantly higher temperature (240°F) in order to kill bacteria and botulism toxins, and this is accomplished with a device called a **pressure canner**.

Most experts agree that it's easier for beginners to start out by canning high-acid foods in the water-bath process. There's less risk involved and the process is simpler. Once you've mastered the steps and are comfortable with water-bath canning, then you might want to expand your horizons by giving pressure canning and low-acid foods a try.

Thankfully, the equipment required for canning is minimal, and—with the exception of the rubber-sealed lid tops for the jars—it's completely reusable year after year. Once you've made the initial investment in your canning equipment, the expense thereafter is very low.

The specific steps for canning vary depending upon a variety of factors—the type of produce (low-acid, high-acid, somewhere in between), the specific recipe, the altitude of your area, and so on; the varying steps are too complex to explore within the confines of this text. But just to get you started, here's a checklist of items that you'll want to have on hand before you begin canning. (Additional items such as dish towels and saucepans are necessary,

Use Mason or Ball canning jars and lids. Don't waste your time by using second-rate jars or lids that are old and cracked or don't quite seal.

Cleanliness is of primary importance in successful canning. Sterilize your equipment, use clean towels, and wash your hands.

but this list only includes items that you might not already have in your kitchen.)

- Jars and lids: You'll want to get your hands on some Mason or Ball canning jars. These are the undisputed leaders in canning jars and are widely accepted as the best choices. You're going to a lot of trouble to can your fruits and vegetables and you don't want to waste your time and effort by utilizing makeshift jars that aren't really suited for canning.

- Funnel: A funnel is a vital piece of canning equipment. How else would you get your tomato sauce into that canning jar?
- Jar lifter: Lifting those canning jars out of boiling water would be impossible without this handy tool.
- Water-bath canner, lid, and rack, or pressure canner, lid, and rack: The heart of the canning process!
- Jar wrench or lid wrench: You'll find this tool useful for removing stubborn lids.
- Magnetic lid lifter: A tool that helps you to maintain the sterility of the lids during the canning process.

In the canning process, the food that is being canned—applesauce, salsa, jam, vegetables, fruit, and so on—is funneled into sterilized jars, leaving the appropriate amount of headspace at the top of the jar (the amount of headspace will vary depending upon what you're canning—recipes will vary). The rims are then wiped clean and the lids and rings are placed on the jars. The jars are placed in the canner and completely submerged in boiling water. The level of water needs to exceed the top of the jars by at least 2 inches, and the jars must be placed on a rack inside the canner. The jars cannot sit on the bottom of the canner—cracking and breakage of the jars could occur. The precise length of time necessary for the boiling process varies depending upon a wide range of criteria.

When the canning process is complete, remove the jars, one-by-one, from the canner (use your jar lifter tool for this). After 24 hours, make sure that the lids on each jar have sealed properly. Look for a slight indentation in the center of each lid—this indicates that the vacuum seal is in place and the jar is correctly sealed. If you find that one of the jars has a top that moves up and down, this indicates that the jar is not properly sealed. In this case, pop the jar in the refrigerator and consume the contents within two weeks—a jar with a broken seal is not safe to keep in an unrefrigerated environment.

For additional information on canning and step-by-step advice, be sure to check out books such as *The Fresh Girl's Guide to Easy Canning and Preserving* by Ana Micka and *The Ball Complete Book of Home Preserving* by Judi Kingry and Lauren Devine. These volumes completely detail the processes of water-bath canning and pressure canning and provide extensive information and recipes.

Safety First!

It's extremely important to carefully follow the necessary steps for canning

Freeze berries first in a single layer on a cookie sheet. Once they're frozen, put them into a storage container.

in order to ensure food safety. Cleanliness is one of the main keys to successful canning. Aim for a scrupulously clean environment, sterilize your equipment, and always use clean towels and wash your hands thoroughly.

Obviously, you must use common sense and caution when consuming your preserved foods. Don't eat food from any jars that were improperly sealed during the canning process, and discard any jars that display mold. You'll also want to avoid any jars in which the contents have changed color or look cloudy, or any jars that have an odd or unpleasant odor. It's always better to be safe than sorry—so be sure to discard any potentially spoiled jars.

Freezing

Freezing your produce actually allows it to retain a higher percentage of nutrients—nutrients that might be otherwise lost in the high heat of the canning process. Additionally, the issues surrounding bacteria and toxins are eliminated because most microorganisms cannot survive in such cold temperatures. So if the idea of canning

Blanch vegetables (boil briefly, then plunge into ice water) before freezing them in order to preserve their color and flavor.

makes you nervous—and it does intimidate some people—then be sure to explore the option of freezing the fruits and vegetables that you've harvested.

Freezing fruits and vegetables doesn't require any special equipment—only a large freezer with plenty of space to store your frozen treasures.

Berries are one of the easiest fruits to freeze; simply place the clean berries on a baking sheet in a single layer and place them in the freezer until completely frozen. Then transfer the berries to a storage container. This method works for raspberries, blackberries, blueberries, strawberries—and it's as easy as can be. You can also effectively freeze grapes and cherries in the same way.

Other types of fruit can require additional steps for freezing; many people choose to peel, core, and slice their apples prior to freezing, as well as treat them with lemon juice to prevent browning. It's possible to simply freeze whole apples, but it's generally considered to be less effective.

Tomatoes are another "easy freezer"—simply remove the cores and place the tomatoes on a baking sheet until they are frozen; no blanching is required. In fact, freezing veggies is fairly straightforward, but you must add in the extra step of blanching the produce prior to freezing it.

Blanching is a simple process in which fruits or vegetables are placed in boiling water for a short period of time, then immediately transferred to ice water in order to stop the cooking process. Blanching is an effective way to preserve the color, flavor, and consistency of your frozen vegetables.

The length of blanching time varies depending on the particular vegetable; green beans, for example, are placed in boiling water for just two to four minutes before being switched to ice water.

Your frozen veggies will keep in the freezer for up to a year, depending on the particular vegetable.

Drying

People have been preserving food for centuries, and dehydrating was one of the earliest methods. Though it may not be as popular in today's kitchens as other forms of food preservation, people are turning once more to the dehydration process, as it boasts a number of advantages. Many consider drying to be less labor-intensive than other forms of food preservation, and you can dry food with equipment that you already have on hand.

Drying doesn't always preserve the flavor of foods as well as other forms of preservation, but is a viable option for many types of fruits, vegetables, and herbs.

Apples are a good choice for dehydration. The best way to dry food is in the oven or in a food dehydrator.

In the drying process, moisture evaporates from the food, which halts the growth of bacteria. If properly dried and properly stored, dehydrated food can last longer in storage than its frozen or canned counterparts—some sources suggest months or years, depending on the type of food.

So if you want to get started in this age-old preservation process, where's the best place to start? Well, as convenient and traditional as it might sound, solar drying your food outdoors is not always a feasible option, especially if you live in a climate where sunshine and low humidity are not prevalent. The humidity level is paramount to the dehydration process; high humidity (increased moisture in the air) is counterintuitive to drying.

As is the case with canning, cleanliness is extremely important when dehydrating.

Oven drying is a viable option that doesn't leave you dependent upon the weather, or you can purchase a food dehydrator (probably the best choice) and dry to your heart's content.

Some vegetables—like dry beans—dry right on the plant, which makes the task of preservation very easy. Simply remove the shells, harvest the dry beans, and store them for future use.

If you made a list of all the foods that you could potentially dry, the list would be extremely lengthy. But for starters, here's a list of good dehydration candidates:

- Bananas
- Apples
- Herbs
- Pears
- Berries
- Grapes
- Plums
- Tomatoes
- Kale
- Meats (for jerky)
- Potatoes

DAIRY PRODUCTS

You can utilize the milk from a family cow as a fluid to drink; process it into butter, yogurt, or cheese; or use it to feed other farm animals including pigs, dogs, and cats. Proper handling of the milk after it has been taken from a dairy cow, doe (goat), or ewe (sheep) is the most important step to making a good end product. Improper handling of milk can lead to off-flavors for drinking and makes other products such as cheese more difficult to process. If you plan to process the milk into other products soon after extraction, then the most important aspect is to remove any foreign matter that may be in it such as hair, flies, dirt, or straw by straining it through a filter pad or cheesecloth.

You can drink raw milk or you can pasteurize it before drinking. Pasteurization heats the milk to destroy microorganisms that may cause disease or spoilage. Raw, unpasteurized milk is safe to drink if it comes from a cow that has tested negative for tuberculosis. Health issues with raw milk often center on unchecked bacterial growth that may occur during the handling period coupled with human resistance, or lack thereof, to any bacteria present in the milk. Proponents of raw milk consumption believe many healthful benefits are lost during the heating process of pasteurization. If you and your cow, sheep, and goat are healthy, there should be little reason to worry about consuming raw milk. You will need to weigh the benefits and risks from your own perspective, but understand that many farm youngsters have been raised for hundreds of years on raw milk. For recipes and step-by-step instructions on making cheese, see *Homemade Cheese*, by Janet Hurst (Voyageur Press, 2011).

You can pasteurize milk in your home by heating it to 150°F for 30 minutes.

Always strain raw milk immediately to remove any hair, dirt, flies and other objectionable matter that may have found its way into the milk bucket.

Options for Milk

If not used right away, your milk should be cooled to refrigerator temperatures as quickly as possible to prevent spoilage. Otherwise, try one of these several methods in which to process it.

Pasteurization

Three methods are used to pasteurize milk in your home: heating to a rolling boil for one second; heating to 170°F for 15 seconds; or heating to 150°F for 30 minutes.

Cream

Raw milk will separate overnight during refrigeration, and the cream will rise to the top of the container. You can skim off the cream and use it in your morning coffee or tea, or you can whip it for the tops of desserts or cereal.

Yogurt

Yogurt is a fermented milk product. It requires the addition of a starter of active yogurt culture to set it in motion. The bacteria in this culture produce lactic acid during the fermentation of lactose. They lower the pH, give yogurt its tart taste, and cause the milk protein to thicken. They also act as a preservative because pathogenic bacteria cannot grow in acidic conditions. Yogurt is easy to make at home.

Incubate the yogurt by putting the jars on a warming plate or in a dehydrator, for about four hours. Then refrigerate.

When yogurt is cool, stir in desired fruit and serve. Yogurt should keep in the refrigerator for seven to ten days.

It's easy to make butter from your cow's or goat's milk. Use an electric churn to speed up the process.

Butter

Butter is easy to make. It requires only cream, a bowl for mixing, and a large jar with a lid. You can use a food processor or an electric churn to take some of the arm work out of it. Let cooled cream warm to room temperature. Then blend or stir constantly until butter clumps together. Drain off remaining buttermilk, fold and press the butter to remove remaining moisture. Rinse in clean, cold water, drain, and add salt, if desired.

Add warm cream to the churn and agitate. It will slowly transform into the first stage of butter. Pour into a cheesecloth to drain the buttermilk that remains.

Remove cloth and butter and place in a clean sink or bowl. Wash with cool water until the water is clear.

Work the butter with your fingers to remove any trapped buttermilk.

If you salt the butter, mix it thoroughly to avoid salt pockets. Roll salted or unsalted butter into balls or spoon it into molds or shallow pans.

You can easily assemble the equipment and ingredients you need to make hard cheese. The process demands steady attention, though, so you'll need a span of uninterrupted time.

Ice Cream

You don't need expensive equipment to make homemade ice cream. You can use a hand-cranked or electric ice cream maker, or you can do it by hand. Two basic types of homemade ice cream are common: one uses eggs for a custard style that is creamier, and one doesn't contain any eggs. The eggs let the fats and water mix together better, which adds richness to the ice cream. They also help it hold up better against melting.

Cheese

Basic cheeses are relatively easy to make and range from soft to hard or young to aged. Farm-produced cheeses are in demand. You can learn how to make many different kinds for your family or to sell under your own label. To make cheese, the solid parts in the milk—the proteins—must be congealed as curd and then separated from the liquid portion, called whey. You will use a bacteria culture called a starter, which is added to the warm milk. When thoroughly mixed, this will start the cheese-making process. To this mixture is added a rennet extract that acts as a curdling agent. As the proteins coagulate, the whey volume increases as the liquid separates out of the curd. It is cut, cooked, and drained, with salt added if desired, before it is cut again and pressed. The excess whey can be fed to pets or other animals such as pigs. Many different cheeses can be produced from cow, sheep, or goat milk.

Cottage Cheese

Making cottage cheese is very similar to making hard cheese. You can use the same steps until you reach the stage where you have to cut the curd; instead, stir it over low heat. The temperature is then raised before draining off the whey. It is then flushed with cold water before salt is added, if desired.

BUTCHERING

The home butchering processes used for different species have many similarities and steps. Depending on whether you use your own live animal or purchase a dressed carcass, some of these steps may have already been done for you.

The process for handling a large animal carcass is similar for beef and dairy cattle, pigs, sheep, or goats; it is the live animal that may pose a challenge. A 100-, 200- or 1,000-pound live animal can vary in attitude and temperament. Any animal sensing a threat will react in unexpected ways. If you choose to work with a live animal, be sure you have sturdy gating and pens, a plan to quickly and safely to restrain and dispatch them, and proper, sturdy, safe equipment that is ready to use. Preparation for your harvest should include a thorough knowledge of the carcass, sharp and clean knives, and meat cutting saws. You must have adequate help available when needed.

You can eliminate the concerns about handling live animals by arranging the purchase of an animal and have it killed at a local meat locker. Then you can retrieve the carcass to cut it up yourself if you have a safe, sanitary, and refrigerated means to transport it.

Regardless of species or size of the animal being butchered, five common steps are involved in ensuring a safe and effective harvest to retain the

If you plan to butcher at home, be sure to plan ahead: have the right equipment, sturdy gates and pens, and adequate help.

most volume and meat quality possible: (1) downing, (2) bleeding, (3) evisceration and skinning, (4) splitting and fabrication, and (5) preservation. The following information assumes the use of a live animal. Purchasing a dressed carcass eliminates three of these steps.

Butchering at home requires some thought before, during, and after the entire process. Planning ahead for this event will minimize mistakes, reduce the chance of injury to yourself or your helpers, and provide you with quality meat products for your family. Accurate information on all aspects of the butchering process, from slaughtering to processing, and the preservation of the meat is useful when you plan to do your own.

A general overview of the butchering process is given here. For detailed, step-by-step instructions for deconstructing carcasses, see *The Complete Book of Butchering, Smoking, Curing, and Sausage Making* (Voyageur Press, 2010), which provides more in-depth information.

Develop a list of all the steps needed, from beginning to end, several days before butchering your animal. Having a list of the equipment and butchering tools required will ensure everything is at hand when needed. Attention must be paid to cleanliness at all times so the meat isn't contaminated during any part of the process. If you're butchering for the first time, review sketches of the skeletal structure, digestive system, and placement of organs, and have some understanding of the circulatory system. This will be useful when you are cutting any large or small animal carcass.

Knives and Other Equipment

You should have all of your equipment, knives, and saws ready for immediate use prior to downing the animal. You should have several knives available,

Four different knives can accomplish most all tasks involved with slaughtering and fabrication of animals for home harvest. These include, from left to right, a 6-inch curved (flexible) knife, a 6-inch straight (stiff) knife, and 8-inch breaking (steak) knife, and a 10-inch breaking knife.

one meat saw, a catch pan for the blood, a metal or plastic tub for the entrails, a pan for the liver, and any other items you deem necessary. All of these items should be thoroughly washed and sanitized before they are used. A pail with soapy water and one with clean warm water should be available to wash your hands, and towels and cloths should also be handy.

A large animal carcass, such as beef, has considerable weight and the equipment you use to suspend it while working on it needs to be stable and strong enough so that it won't tip, buckle, or break while in use. Even smaller carcasses, such as a pig, may have weights that challenge the equipment you might have.

Any carcass that falls to the floor or onto the ground, whether large or small, will be difficult to lift if no alternative method is available. Having to lift again a carcass that has fallen from its holding can cause delays in processing the meat, which can lead to spoilage. Also, potential damage from bruising of the muscles or possible contamination of the carcass by dirt, manure, or any foreign substances it comes in contact with may occur.

Knives will be needed from the start of the butchering process until your last cut is made. The number or style of knives you use may range from a small hand knife to a large, sturdy butchering knife, and a variety in between.

Different knives are available to make certain cuts easier and more precise while other, larger knives have many advantages for cutting up larger pieces of the carcass. One simple rule is that sharp knives always work best. However, they also carry safety concerns when using them.

QUICK TIPS FOR KNIFE USE

- Always use a sharp knife when cutting meat.
- Never hide the knife under your arm or under a piece of meat.
- Keep knives visible.
- Always keep the knife point down.
- Always cut down toward the cutting surface and away from your body.

Personal safety for you and anyone working with you is of prime importance when handling live animals, slaughtering them, and cutting up the carcasses. Being injured by live animals can have devastating consequences. Similarly, you can be injured by unstable or inappropriate equipment, whether it is being used for slaughter or for food processing. Knife injuries can occur quickly and unexpectedly, and in severe cases, they may be life threatening. Common sense, caution, and alertness to potential dangers will help avoid serious injury.

Knives and Saws

Knives are typically available with wooden or plastic handles, come with flexible or stiff blades, and are made in many sizes and shapes.

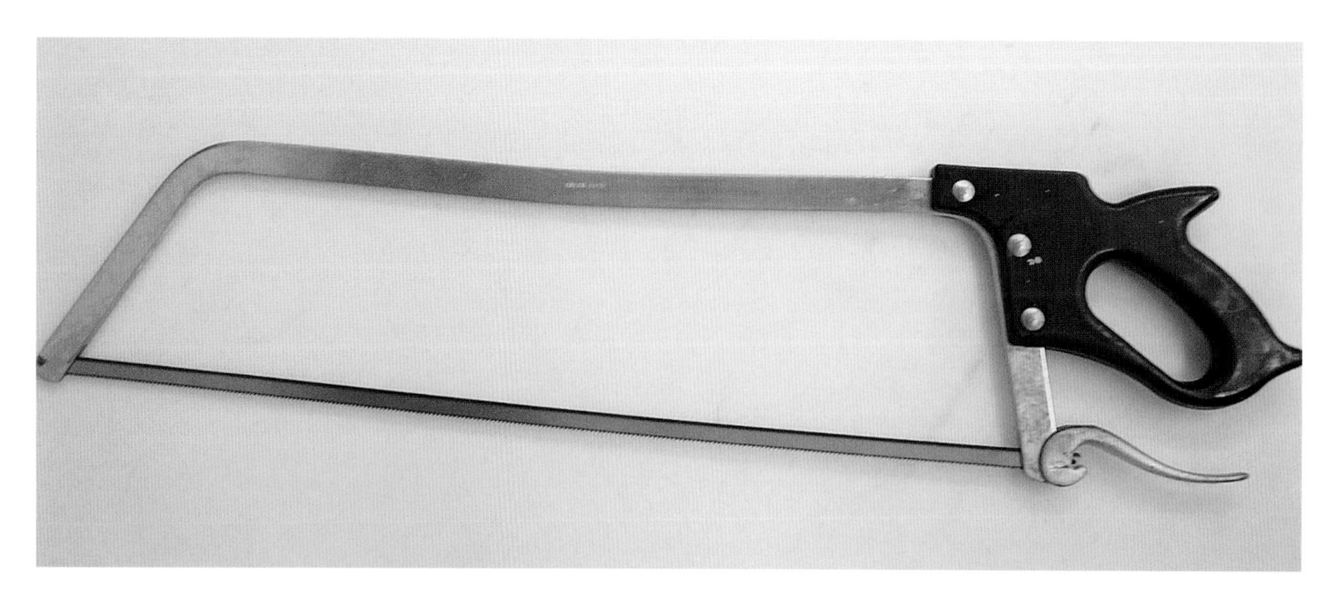

Meat saws are used to cut through bones or other areas of the carcass that may be less accessible for knife use. Most meat saws range from 12 to 25 inches in blade length.

Knives that are not sharp pose a safety hazard. They can slip as more effort is required to pass the knife through the muscle or bone. If your knives are not easy to clean and kept sanitary, they may harbor harmful microorganisms that can affect the quality of the meat and possibly your health.

Meat saws are used to cut through bones or to sever portions of large carcasses into smaller, more manageable ones. Most meat saws are between 12 to 25 inches in length with a serrated blade. Blades should be complete, and those that have developed rust spots or have chipped or missing teeth should not be used. Any meat saw should be thoroughly washed and sanitized before use, paying particular attention to the area where the handle attaches to the metal frame.

Other Items

Gloves are one of the best protective items you can use. A butcher glove is designed to be worn on the free hand. They come in several sizes and are easy to wash. A heavy mesh glove is made of solid stainless steel rings to protect hands against cuts, slashes, and laceration hazards but may not entirely stop punctures. Before and after each use, you should thoroughly clean, sanitize, and dry them.

Electric knives can be used in place of standard knives. Electric knives may be easier to use to carve or fillet different cuts of meat, particularly if handling heavy portions is a concern.

Aprons made from leather, Naugahyde, heavy canvass, or rubber can be a protection from injury or keep your clothes from becoming soiled or bloody during the slaughtering process of large animals.

Cutting Surfaces

Cutting surfaces or cutting boards should be made of material that is easy to clean and fairly soft. Natural wood or synthetic materials such as soft

A heavy mesh butcher's cutting glove is worn on the off-knife-holding hand and is designed to protect your hands against cuts, slashes, and punctures from your knives. Always wash and disinfect the glove before and after use and dry properly. Gloves of this type are sold in different sizes and can be used on either hand.

QUICK TIPS FOR FOOD SAFETY

- Always wash hands with soap and water before handling meat or beginning work. Rewash between tasks, after sneezing, using toilet facilities, or handling materials not part of your processing work.
- Before and after use, thoroughly clean all equipment, knives, utensils, thermometers, bowls, and anything else used to cut or store meat. Clean and sanitize all surfaces that will be used.
- Keep raw meat separate from other foods. Avoid cross contamination between pieces of raw and processed meats. Avoid mixing of fluids and juices from other cuts or vegetables to be used.
- Keep meat below 40°F during processing.
- Monitor temperatures at all stages of your processing.

polyethylene are good cutting surfaces. Avoid using glass, ceramic, metal, marble or any other hard surface material for cutting meat because these can have a damaging effect on knife blades and edges.

Cutting surfaces can provide an ideal area for cross-contamination of food products, a major food safety concern. Bacteria transferred from knives to cutting surfaces or cutting boards to other foods can lead to food poisoning. Always clean and sanitize the surface you use for cutting meat before and after each use.

Sanitation

Strict sanitation is critically important at all times when handling raw meat and must be maintained to prevent bacterial contamination and food-borne illnesses. It is essential to handle raw meat in a safe manner that reduces the risk of bacterial growth. No meat product is completely sterile, but using proper procedures will minimize your risks. The most basic sanitation procedure involves using and maintaining clean surfaces before and after processing sausages. It is easy to remember the 3 Cs of sanitation; keep it clean, cold, and covered.

Keep It Clean. Wash all surfaces that you use with a diluted chlorine bleach solution of 10 parts water and 1 part bleach, as well as antibacterial soap. Nothing will replace vigorous scrubbing of the surface area with these products. This solution will remove any grease or unwanted contaminants from the preparation area. Keep the area free of objects or materials that do not relate to meat preparation or that will be used later. Utensils and your hands should be thoroughly washed before beginning. Be sure to remove any rings, jewelry, or other metal objects from your hands, ears, or other exposed body parts.

Restraining large animals is the best way to assure a clean kill and allows you to properly place the compression gun or rifle used. A stunning gun renders the animal unconscious so that it feels no further pain but allows for a more complete bleed because the heart is still pumping.

Keep It Cold. Bacteria grow best in temperatures between 40°F and 140°F. If you are cooking or cooling meat for cooked sausages, be sure your product passes through this range quickly because meat can be kept safe when it is cold or hot, but not in between. The meat you process should pass through this temperature range, whether being cooked or cooled, within four hours, but preferably less. This includes any butchering time involved.

Keep It Covered. Meats, carcasses, and wholesale or retail cuts should be covered during any time you are not working on them. Your processing equipment should be properly stored in between use as well as any area used where butchering is done. Maintain screens, barriers, or traps to keep out vermin and reduce access for flies and insects.

The Butchering Process

Learn the steps in the butchering process before you begin, so you can move easily from one to the other. Your goals will be to process the meat quickly in a way that avoids contamination and to avoid injury to yourself and any helpers. The quicker the process, the sooner the meat can be safely stored.

Downing

The location where you plan to carry out the kill should be properly equipped for the job. A shed or building that is free from dust or outside elements can provide a good place for the initial stages.

A proper set of butchering tools includes sticking knives, skinning knives, boning knives, butcher knives, steel sharpener, meat saws, and meat hooks. Other useful items include thermometers, meat grinder, hand wash tubs, clean dry towels, soap, and vats with hot and cold water.

If you're working with a live animal, you will be faced with the decision of how to put the animal down so that you can begin the first of the harvesting processes: the sticking of the jugular vein to facilitate bleeding of the carcass. An animal can be dispatched in several ways, and none of them are for the faint of heart. A misapplied stun or gunshot will result in a frantic animal that's both harder to approach for a second attempt and has an increased heart rate, which releases adrenaline and lowers the meat quality of your carcass.

Stun or shoot the animal in the forehead, at a point where imaginary lines from each eye to the opposite horn root, or pole, cross. If the animal does not have horns, imagine where they would be if they did.

To raise the animal off the ground or floor, tightly chain the hind legs together between the hock and feet and lift it with a winch or loader. Be careful to avoid injury to yourself because once an animal is stunned or shot, it may exhibit involuntary body reflexes with the legs kicking or thrashing. They usually subside within the next few minutes.

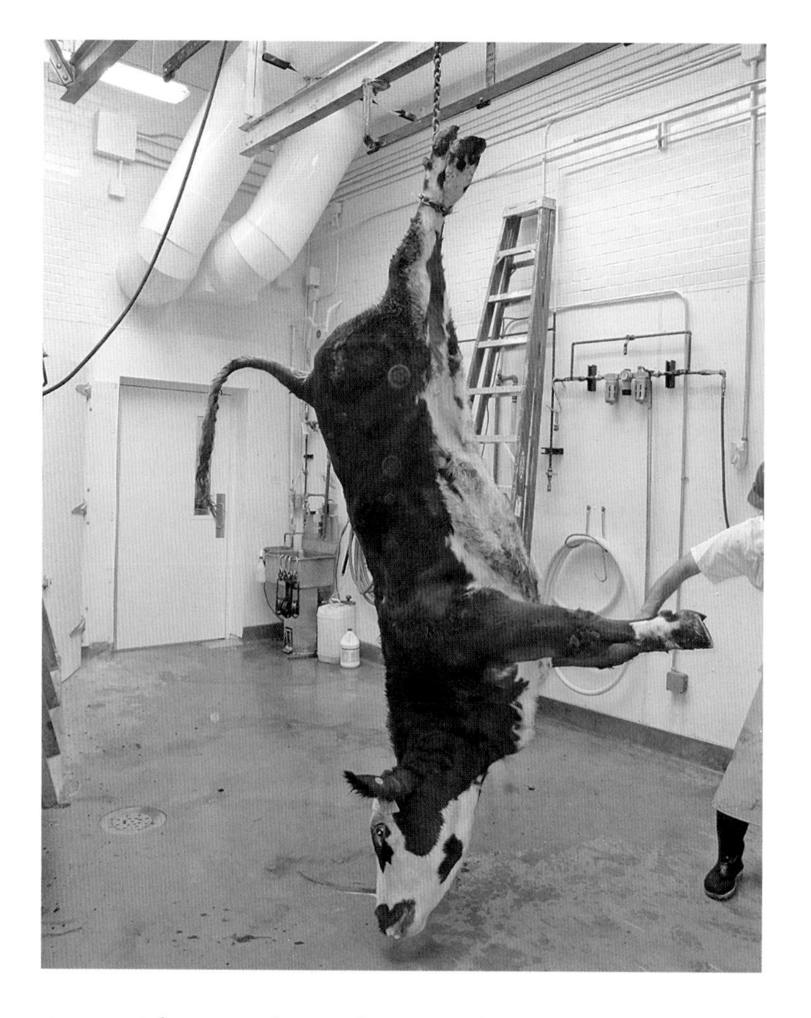

A major consideration about where you down the animal is how you will raise it off the floor or platform where it is standing. The weight of the animal should be a consideration in how you approach this procedure as well as the height of any ceiling present. You can use a shed or even lift the animal to an open area if you have the machine to handle it properly. Remember that once the animal is stunned or shot you need to begin work to bleed it quickly after raising it in the air.

Your work should proceed in an area that is clean and free from dust, dirt, insects, and anything that might contaminate the carcass once it's opened. Dripping blood will quickly attract insects and flies that can lay eggs in a very short period of time. For these reasons, it is best if the sticking and evisceration is done in an enclosed area.

It is important to handle any animal calmly before slaughter. Physical stress before slaughter can particularly alter the pH levels in the muscles and be a cause of dark-colored meat. Keeping the animal calm and quiet is the best prevention for optimum meat quality.

Bleeding

The process of bleeding an animal—whether large, such as beef, or smaller, such as pigs, goats, and sheep—is similar. Once the animal is unconscious, you can wrap a chain around the end of the canon bone above each rear ankle. These bones are strong and will allow you to raise the carcass. The standard method for sticking is to make an incision between the brisket and jaw. Peel the skin apart to expose the carotid arteries and jugular vein and sever them with your knife. Catch the blood in a large tub, vat, or barrel. The blood volume may vary between animals but will generally be between 6 to 8 percent of the live weight. A good stick will remove about 50 percent of the total blood in the carcass.

The head of a large animal should be one of the first parts removed to lessen its weight, to aid in bleeding, and to provide easier access to the carcass. Begin by making a cut from the poll at the top of the head down the center of the nose tone nostril and down to the jaw. You can skin out one side of the face, peeling the skin back as you go, before skinning out the other side. Grasp the bottom jaw with your free hand, pulling upwards so the poll bends back and cut through the Adam's apple and the joint at the base of the skull.

In the past, some families made use of the brain. However, because of the development of links between Bovine Spongiform Encephalophy (BSE) found in infected cattle and variant Creutzfeldt-Jakob Disease (vCJD) in humans, you are strongly advised **not** to eat any part of the brain, spinal column, and other parts of the nervous system. Discarding the head, except for the cheek meat and tongue, and any remnants of the spinal cord is the safest route. Also, you should not feed the brain, spinal cord, blood, or other nervous system parts to other livestock or chickens. This will reduce the potential for any transference of infective agents from one animal to another.

Removing Legs

Your next step is to remove the legs to prevent possible contamination of the inner carcass with manure and dirt dropped from the hooves. You can do this while the animal is suspended or if you have a skinning cradle, to lay the carcass out and remove them while the carcass is on its back.

Use the tip of your knife to open the skin, starting with a circle cut around the backside of the front leg near the dewclaw. Cut a line up the foreshank until you reach the elbow and then continue across to the midline of the brisket. Peel back the skin to expose the entire leg bone.

To remove the foreleg, cut across the shank to sever the tendon, which will release the tension on the lower part of the leg. Next, cut through the flat joint, which is about 1 inch below the knee joint. If it is too difficult to cut with your knife, use your meat saw. Then make the same cuts on the other foreshank.

The procedure for removing the hind legs is almost identical, except you will be making your initial cut up the inside of the hind leg and across to a midline point directly below the anus. In removing the hindshank be sure to make your cut below the point where the tendon anchors itself to the joint. This will allow you to hang the carcass by the tendons, which are strong enough to hold the weight. However, to do this, the tendons must still be attached and intact.

Skinning

Remove the hide and any manure, dirt, and hair. To open the hide, you can start at either end, and this is easier if the carcass is on its back. Pull the hide upward as you make a cut from the throat to the anus, following an imaginary midline of the carcass. Pull the hide toward you to prevent cutting into the carcass or through the abdominal wall. Next, firmly grasp the hide and use your skinning knife to make long, smooth strokes to separate and peel the skin from the carcass.

After removing both sides of the hide as far as possible while the carcass in lying on its back, you can open the brisket. To do this, use your knife to cut through the fat and muscle covering it. When the brisket bone is exposed you can use a saw to open it. You can separate the esophagus and trachea now unless you did it earlier before the carcass was suspended. Once the brisket is opened, you can raise the carcass.

For goats and sheep, to separate the skin and fleece from the body, grasp the pelt at the cut, make a fist with your free hand and slide it forward separating the skin from the body. Push your fist against the pelt and not the carcass as you are loosening it. By repeating this motion you will loosen the skin without needing to use a knife. This will eliminate cuts and bruises to the body of the carcass. Always have clean hands when loosening the pelt to avoid contaminating the carcass with wool and dirt from the fleece.

For pigs, you can skin the carcass while it is still lying on its side or back and finish it when it is fully suspended. Be sure that if the smaller animal is lying on the ground that there is a clean canvas or similar sheeting between the carcass and the floor.

To lift the carcass, use hooks that should be attached to the hind leg tendons and lift so that the legs spread apart when suspended. Raise the carcass to a level that is comfortable to work with and is clear of the floor space.

Because of a large animal's carcass length, it will be easier to split the pelvic bone, or the aitch bone, before it is fully suspended and while still at a convenient height. It will also be easier to cut the anus loose, remove the tail and the hide from the rump and rear quarters before lifting it. If it is a male carcass, remove the pizzle first by cutting it loose from the belly and back to the pelvic junction where it originates.

Cut through the muscles and membranes at a center point in the pelvis to expose the aitch bone; using your saw to cut it in half. Loosen the anus by cutting completely around it, severing all connecting tissue. Be careful not to cut into the intestine. When it is loose, tightly tie the end shut with a clean cord or clean heavy string and let it slide into the body where it can be reached from the belly cavity later. You can remove the pizzle with the anus. Remove the tail by severing the two joints where it attaches adjacent the body and cut the skin completely around its base. You should now be able to pull out the tail.

Remove the remaining hide by starting at the top and running your skinning knife down along the carcass. The weight of the hide will help separate it from the carcass.

Rinse the carcass with clean, lukewarm water before opening the body cavity. This will remove any dirt, hair or wool, or other foreign materials that may have attached to it. The carcass is now ready to cool if you choose.

Evisceration

To open the body cavity, start at the point where you cut through to the aitch bone. Slice an opening large enough to insert your knife, handle first, into the cavity and position the blade upward and outward. This allows you to protect the intestines and rumen with your fist. You do not want the blade to cut into the intestine or rumen or you will contaminate the carcass with fecal and rumen materials. Since you have already opened the brisket, you should make one continuous cut from the top down to the brisket opening.

As you slice down the belly, part of the viscera will spill outward but will still be held by membranes that hold the anus, intestines, liver,

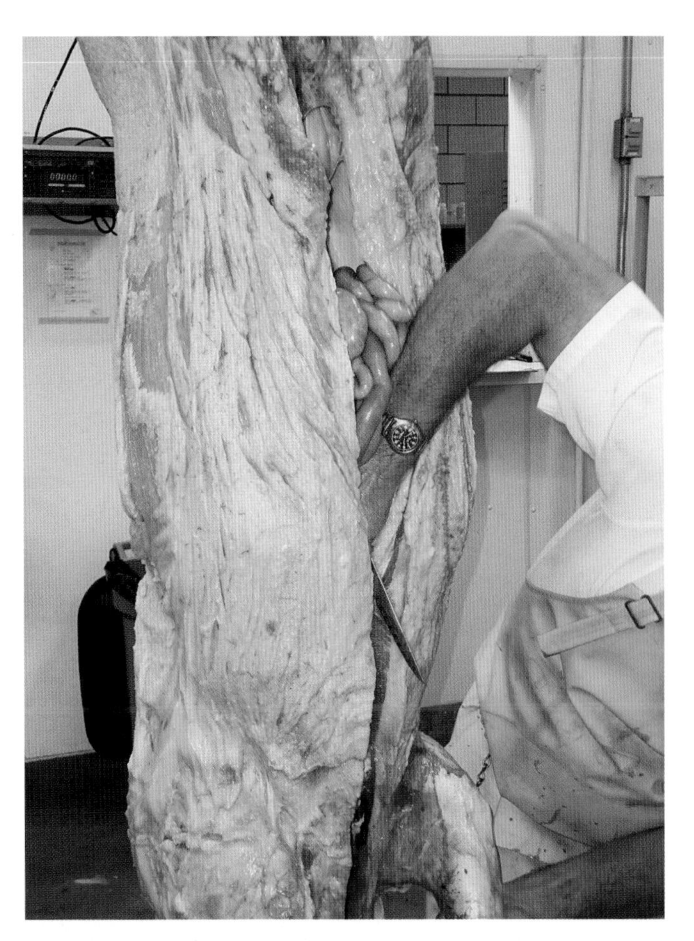

Open the abdomen by starting at a midline point at the pelvis. Once you carefully slice an opening large enough to insert your hand and knife, turn the blade outward while holding the heel with your hand on the inside. Use your hand to hold back the intestines as you slice down the midline in a smooth, continuous motion until reaching the opening of the thoracic cavity.

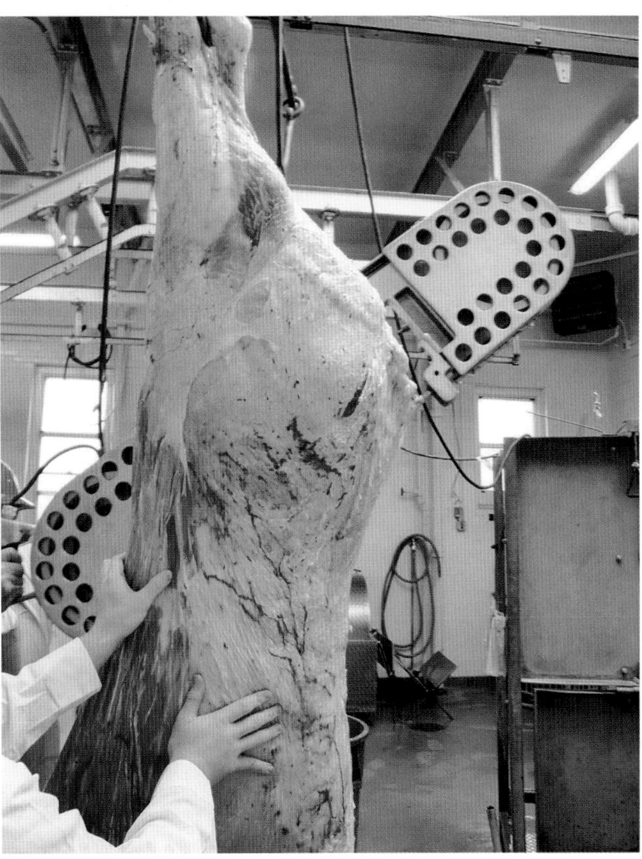

You can split a carcass using a handsaw or an electric meat saw. Start at the aitchbone and carefully cut down the middle of the spine and down the center of the spinal column. If done correctly, the backbone will be split in half, and the loin eye muscle will not be scored or cut into.

and bladder to the inside body cavity. By this point, you should have placed a tub beneath the carcass to catch the viscera after it is cut loose and to collect any blood still draining from the carcass.

With the belly completely open, sever the fat and membranes that still hold the viscera. Start at the top and cut the ureters that hold the kidneys. These can be removed later. You can loosen the liver with your hands and then sever it from the backbone with your knife. Set it in a separate pan for later inspection.

As you loosen more connective membranes, the weight of the viscera will cause it to drop outward. As it does, pull the loosened esophagus up through the diaphragm. This should allow everything to fall freely into your collecting tub.

Chilling the carcass makes it easier to cut up the various parts as the fat within the meat and the muscles become firm. Otherwise you can proceed to cut up the entire carcass.

Splitting

With all the internal organs and intestines removed, you are ready to split the carcass in half. You can use your hand meat saw or an electric meat saw. Begin at the top and slowly make your cut in the exact center of the spine. Continue down until each half is free. Wash the carcass inside and out with cold or lukewarm water to remove any remaining blood, tissue, or foreign material. Splitting the carcass in half allows for easier handling and fabrication. When you have finished with the carcass, inspect the liver and other internal organs to assess their health. A healthy looking liver, pink or salmon-colored and free of lesions or dark spots indicates a healthy animal. The liver and heart may be cooled and used later in sausage making. Generally the intestines from cattle are too large to be very useful as casings in making sausages. Pig intestines are better for casings.

If you plan to use pig intestines for sausage casings, they will have to be cleaned and rinsed with a salt solution several times. The easiest way to do this is to turn them inside out after cutting them into several lengths and scrape off the mucous coating. Generally, the small intestines are used for sausages, so tie off the large intestine, sever it from the small intestines, and discard it.

Fabrication

Fabrication is the term used for cutting the large carcass down into smaller portions for eating and preserving. Larger animals yield more meat than smaller ones. For example, a beef animal typically will have a 60 percent dressing weight after the head, skin and viscera are removed. If it weighs 1,000 pounds, its carcass will weigh about 600 pounds or 300 pounds per side. It will typically yield at 60 percent or more (with bones included in cuts), or about 360 pounds to process.

Market goats and lambs generally weigh about 100 pounds when sold and with a 50 percent dressing average will yield a 50-pound carcass. With an average bone-out of about 30 percent, you can expect a yield of about 15 pounds from them.

A 200-pound pig, with a typical 75 percent dressing weight, will yield a carcass of about 150 pounds, or 75 pounds per each side. Using an average cutout rate of 60 percent for pigs, this 150-pound carcass will yield about 90 pounds of meat for your use. The other 60 pounds will include fat trim, bones, and skin.

Beef

Each beef carcass can be divided into quarters: the two forequarters and two hindquarters. Each forequarter consists of five major cuts; chuck, rib, brisket, plate, and shank. The hindquarter contains the most valuable retail cuts including the round, loin, and flank.

The chuck is the largest cut on the beef animal and the two (right and left) will account for about 25 percent of the carcass weight. Although the chuck contains much connective tissue and is often made into roasts, a considerable amount of lean trim here can be used and several minor cuts can be used in various dishes or sausages.

The foreshank and brisket are considered rough cuts but make up about one-quarter of the total carcass weight, about half of which can be utilized. The plate of the forequarter is the lowest part of the ribs but does not include part of the brisket.

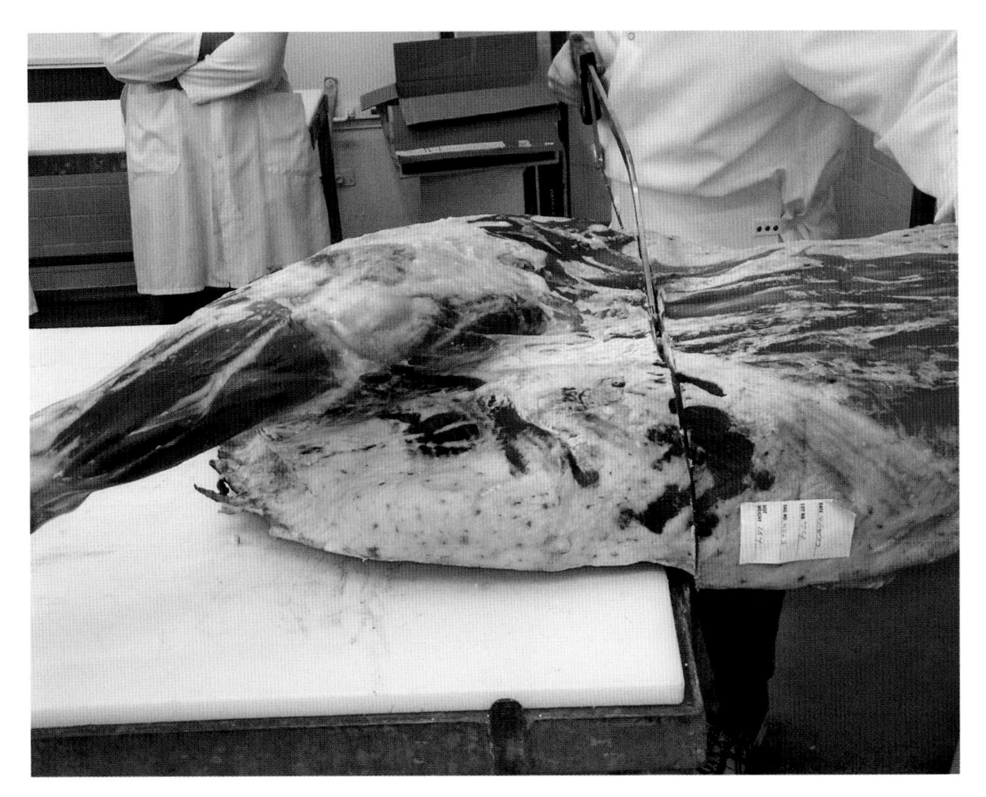

To separate the forequarter from the ribs, make a perpendicular cut to the shoulder between the fifth and sixth ribs. The shoulder, foreshank, and brisket can be set aside until later.

If you don't want to turn your cow into a roast yourself, you can find a local company to do it. Some even have mobile operations and can come to your farm.

The hindquarter contains three cuts that compose about half of the carcass weight; the round, loin, and flank. As with other parts of the carcass, there are several different ways to break down the hindquarter into cuts for your home use.

The flank is used mainly for trim and can be made into ground beef or used for sausage.

The round is fairly easy to cut up. The name of each cut is derived from their position when the round is laid out on a table. The top round is also called the inside round because in its natural position, it would be on the interior side of the live animal. The outside round is also called the bottom round because that is its position when it is placed on the table for cutting; it is on the bottom. The eye is located between the bottom round and the top round. The sirloin tip is that portion that is in front of the femur, or thigh bone, in the standing animal and is composed of four muscles.

The rump is considered as part of the round and can make up about four percent of the carcass weight.

The whole loin is composed of two parts, the sirloin and shortloin. The steak yields from the sirloin are about five percent of the carcass weight while the shortloin will be about seven percent.

Goats/Sheep

There are five primal cuts which begin the process of deconstructing the sheep, goat, or lamb carcass: leg, loin, rack (ribs), shoulder, and foreshank.

The goat and sheep carcass is roughly cut in half and divided into two parts: the foresaddle and the hindsaddle. The foresaddle consists of the

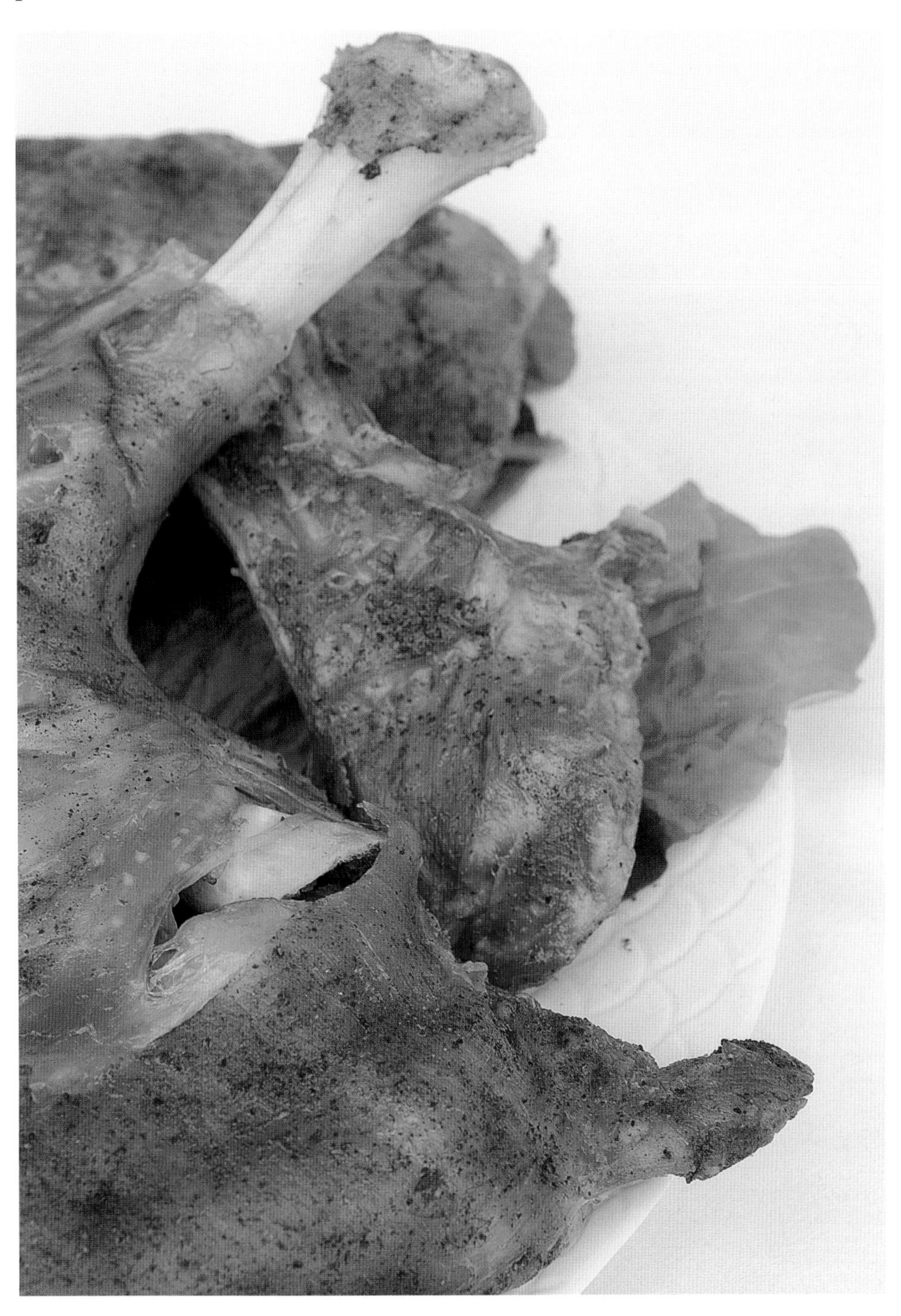

Goat meat is becoming easier to find at specialty markets and at restaurants.

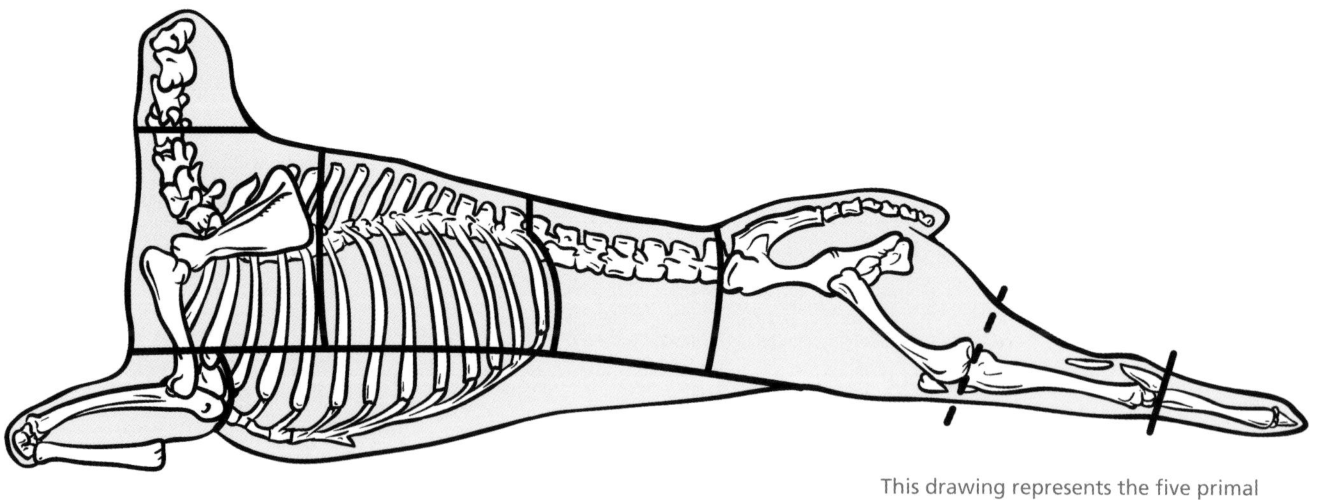

This drawing represents the five primal cuts of a goat or lamb. They are shown in relation to the skeleton, shoulder, rack, loin, leg, foreshank, and brisket.

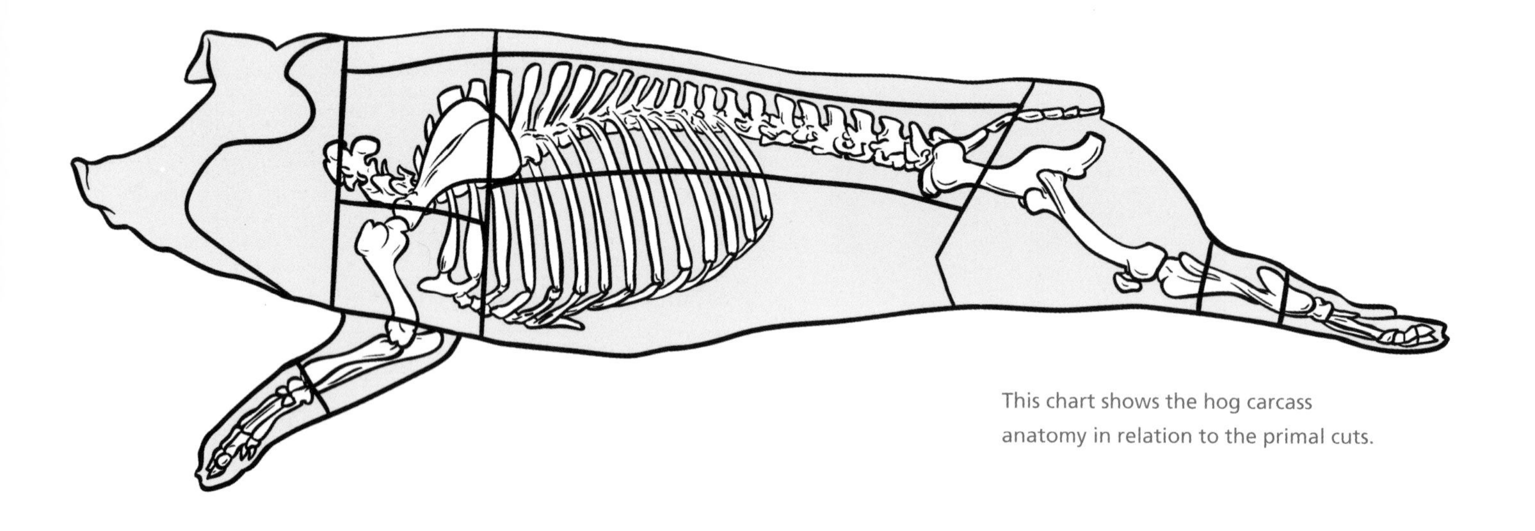

This chart shows the hog carcass anatomy in relation to the primal cuts.

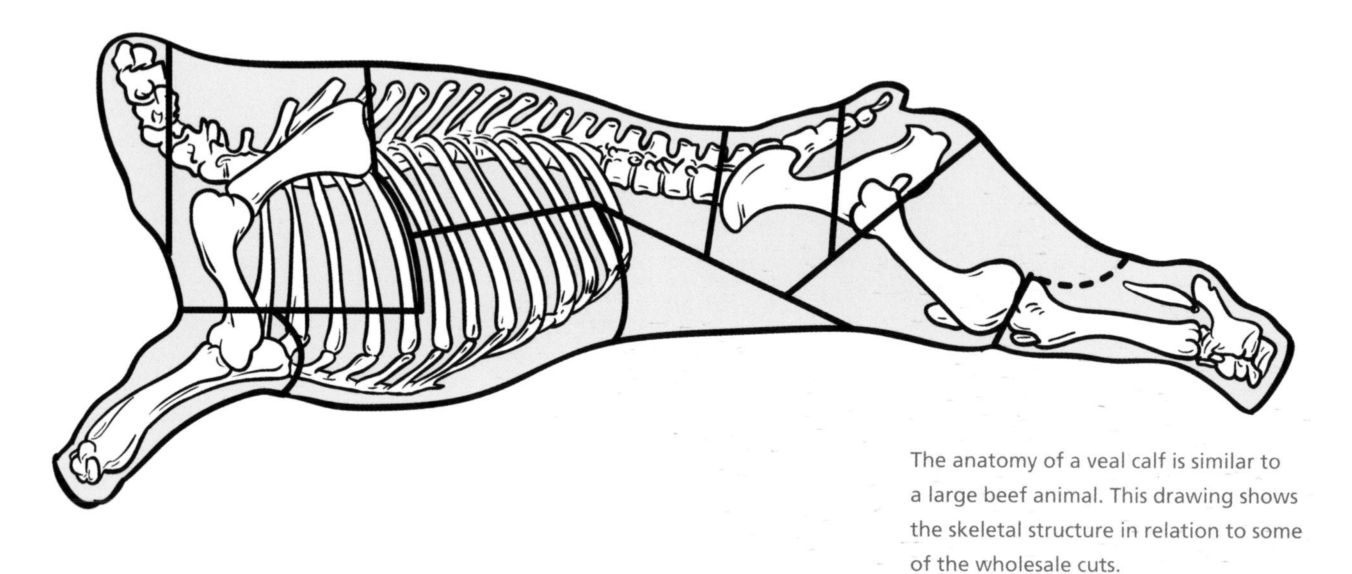

The anatomy of a veal calf is similar to a large beef animal. This drawing shows the skeletal structure in relation to some of the wholesale cuts.

shoulder, rack, foreshank, and breast. The hindsaddle comprises the loin, leg, and flank.

Slightly larger than the rear of the carcass, the foresaddle contains four major portions that can be further reduced.

The shoulder is the largest cut in the foresaddle and contains a number of bones, which makes it more difficult to carve and slice. It is often called a square-cut because it fits the dimensions of a square.

The hindsaddle contains the most valuable cuts of the sheep or lamb. These include the loin, leg, and flank.

The leg is the largest cut and represents about one-third of the lamb carcass. When you cut the leg from the loin, the sirloin will be included with it. Although typically referred to as a leg of lamb, this term also implies that it is the whole leg with the sirloin still intact.

The loin is the most valuable cut in the carcass because it contains the tenderest muscles. The loin and rack (ribs) often compete for the higher price because there are so few of them from each animal.

Several pieces are referred to as rough cuts that include the flank of the hindsaddle and the foreshank and breast of the foresaddle. These typically are more fat than muscle and are usually trimmed as much as possible and used for grinding into patties.

Pigs

The largest part of a pig carcass is the ham, which can be about 23 percent of the carcass. The side or belly and the loin areas represent about 15 percent each. The picnic and Boston butt are each about 10 percent, and the miscellaneous portions including the jowl, feet, neck bones, skin, fat, bone, and shrink account for about 25 percent of the carcass weight. The five major sections in a pig carcass include the picnic shoulder, Boston butt, loin, ham, and belly or sides.

The picnic shoulder includes the upper front leg above the knee. This cut lies just below the Boston butt and contains a higher level of fat than the other cuts, but makes it a flavorful and tender portion.

The Boston butt, also called the shoulder butt, is often a better cut than the picnic shoulder. It lies at the upper portion of the shoulder from the top to the plate to make the backbone.

The pork loin cuts and chops are located directly behind the Boston butt and include a portion of the shoulder blade bone. The loin includes most of the ribs and backbone all the way to the hipbone at the rear. The loin constitutes a long strip that contains the top section of the rib and, when these are trimmed away, the result is a very tender boneless pork loin.

Hams make up about one-quarter of the carcass weight and come from the rear leg area. This area is a prime cut because it contains little connective tissue to make it flavorful whether it is cooked, cured, or smoked.

Pork loin comes from the meat on a pig's back, next to the ribs, and is perhaps the tenderest part.

The pork belly or sides are where the bacon and spareribs are cut out. They are located below the loin on each side. They contain a lot of fat with streaks of lean meat from which bacon is made.

The miscellaneous portion of the carcass includes the jowl, pig's feet, tail, neck bones, skin, and fat.

Rabbits and Small Game

Domestic rabbits are among the most efficient meat producers of all animals. They can reach market weights of 4 to 5 pounds by eight weeks of age while typically converting 2.5 pounds of feed into 1 pound of gain.

After stunning the rabbit, make an incision at the rear hocks between the leg bone and the tendon to suspend the carcass by hooks through the hocks.

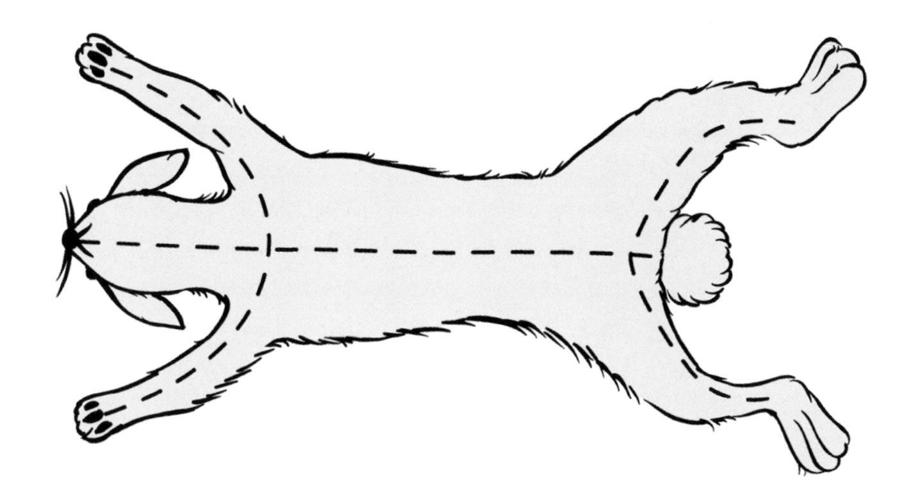

Small furred game can be skinned in one of two methods: open or cased skin. The cut marks on the rabbit show an open-pelt method.

Grasp the hide with both hands and pull in opposite directions. Keep pulling until all the legs are skinned up to the feet. Then cut off the head, feet, and tail. If you did not field-dress the rabbit before skinning, slit the underside from the vent to the neck, and then remove all internal organs.

To skin the rabbit make a cut on the rear side of the loose leg, slicing to the base of the tail and then back up the suspended leg. Pull the pelt away from the muscle and down the carcass. They should separate without needing to make any further skin cuts.

After the hide is removed, you can open the abdomen by cutting the mid-line of the belly from the anus to the rib cage, making sure not to puncture the intestines. Pull out the viscera before cutting off the other rear leg. Inspect the liver for any white, yellow, or other spotting that could indicate an unhealthy animal. Wash and rinse the carcass thoroughly with cold water.

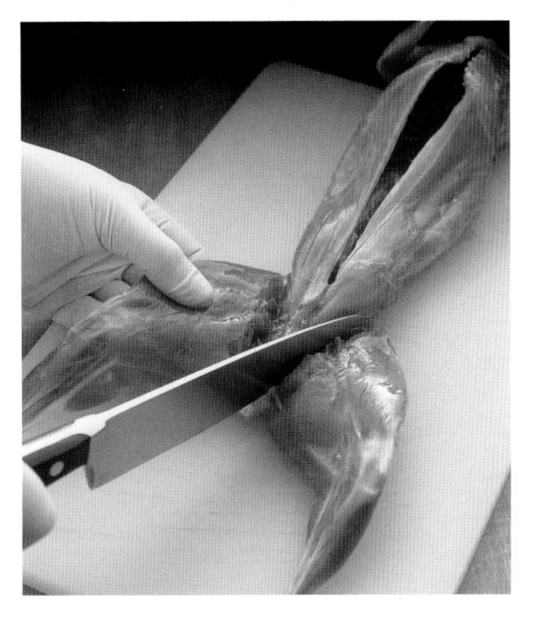

To portion a rabbit, begin by cutting into the rear leg at a point near the backbone. When you come to the leg bone, stop cutting. This portioning method works with squirrels, rabbits, hares, and raccoons.

Then cut the carcass into three sections: forequarters, loin, and rear quarters. The front legs can be split at the shoulders by cutting through the shoulder joints and severing them. The loin can be portioned by cutting off the shoulder and at the hip joint. The ribs can be trimmed because they contain very little meat but can be used for soup stock. The rear legs contain the most meat, followed by the loin and then the front legs.

Poultry and Other Fowl

There is more than one way to butcher chickens, and you can make the process as simple as you wish. One procedure is explained here. Processing poultry requires four basic steps, which should be done in separate areas to prevent contamination:

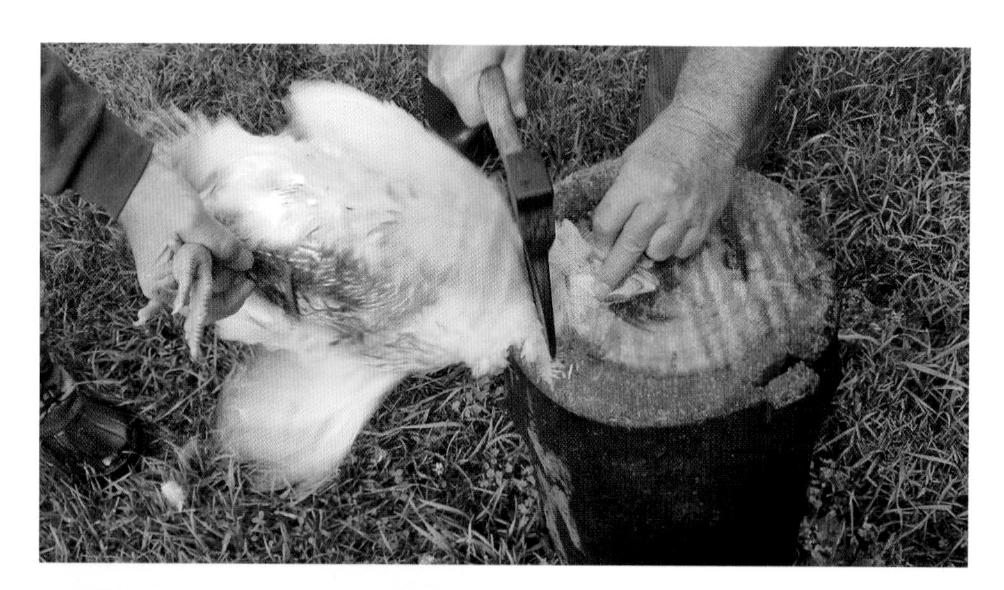

To scald the carcass, hold the feet and gently dip it into the hot water. Hold for five seconds before pulling out. Then dip again while slowly moving it from side to side, completely submerging the feathers. Do not overscald the bird.

Begin by grasping the chicken's feet and hold it upside down to immobilize it until you place its head on the block. Stretch the neck by pulling on the legs as the head is held between two nails pounded1 to 2 inches apart into the block. One swift chop should be sufficient to sever the head from the body. Be sure to keep your hand and fingers clear of the hatchet.

1. Slaughtering
2. Scalding and plucking feathers
3. Eviscerating (cutting open and removing the internal organs) and washing
4. Chilling and packaging

Arrange your work area prior to starting to help move the process along swiftly and safely. If properly done, your processing can be a satisfying experience.

In the most basic operation, you will need knives for eviscerating and cutting, an axe and chopping block for removing the heads, several 5-gallon pails, scalding tub, heating coil, propane tank, canvass or tarpaulin, and a sturdy table. Your chopping block will work best if you pound two large nails into it about 1 or 2 inches apart, depending on the size of the birds.

The area you use for processing should be clean, have plenty of water available, and be as free from flies and insects as possible. Working early in the morning is often a good idea if you expect flies and insects to be a problem later in the day. Scrub tables with soap, water, and a diluted chlorine solution prior to use. Otherwise, use a disposable plastic cover. Sharpen and sanitize all knives before starting. Keep them in a clean and accessible area.

You can use galvanized or plastic garbage cans or pails to hold the cooling water. Be sure these containers have been thoroughly washed, sanitized, and rinsed with clean water each time they are refilled with carcasses to be chilled. Set up similar cans, pails, or plastic-lined boxes to use for feathers and unwanted body parts. While butchering, keep a water thermometer handy for checking the scalding water, which should be kept between 120 to 160°F.

When scalding is complete, begin to remove the feathers by pulling on them. You can place the feathers on the tarpaulin for later disposal. The large feathers can be removed quickly. If needed, you can quickly dip the carcass again to loosen the rest.

You may decide not to cut up the carcasses until all of the butchering, feather plucking, and viscera removal is completed for all the birds. If you wait to cut up the birds, you need to place the eviscerated carcasses in cold water to remove the body heat of each bird as quickly as possible.

Keep your packing materials close by. Have plastic-lined boxes or portable coolers filled with ice where you can cool the eviscerated carcasses quickly.

The method you use to slaughter the bird may involve an axe. Using an axe will require some coordination with a heavy blade, wood block, and an agitated bird that may not wish to cooperate. If you feel you can hold the feet and legs steady while chopping off the head, you may not need to use any other restraints. Once the head is chopped off, you will need to hold the bird by the legs for a few moments until its reflexes stop.

Start by holding the bird tightly by its legs and immerse it neck first into the scald water. It is important to get enough water into the feathers. Move the bird up and down and from side to side to get an even and thorough scalding, which will make the feathers easier to remove. Repeated dips may be needed, but don't overdo them to prevent burning.

One simple rule to follow when scalding is that the higher the temperature of the water is, the less immersion needed. You can avoid over-scalding by following the temperature and time factors. Over-scalding causes the skin to tear and discolor and gives the bird a cooked appearance; the carcass will lack bloom and turn brown rapidly or bright red when frozen.

Hot scalding, with water temperatures above 155°F, is an easy, quick method to remove feathers. Feathers from waterfowl, ducks, and geese are more difficult to remove and can be scalded at higher temperatures of between 160 and 170°F for one to two minutes. Waterfowl have natural water-repellant oils in their feathers. You can add detergent to the scald water used on waterfowl to help the water penetrate through the feathers.

Picking off the feathers, or plucking feathers, is the next step. Birds should be plucked immediately after scalding. You can lay them on a canvas or tarpaulin, or suspend them by their legs to do this. There is no one correct way to remove the feathers. All feathers need to be removed. Chickens and other domestic and wild fowl have pin feathers which are tiny, immature feathers lying below the surface of the larger feathers. They are more difficult to remove because of their size, and you can use a pinning knife or dull knife to gently scrape or pluck them off.

Unless you immediately proceed to cutting up the carcass, put the carcasses in a cold-water bath with temperatures between 32 and 36°F. Birds should never be frozen before being chilled down because the meat will be less tender later as the muscle fibers slide and lock together. Placing the birds in an ice slush will rapidly cool them.

To make your work easier, there is a proper order for evisceration, which is cutting open and removing the internal organs from the body cavity, plus the removal of the head and feet.

Remove the head and neck first. The head will have been removed if you used an axe and chopping block earlier. If the bird was suspended by its feet and you simply cut its throat, you will now need to remove the head and neck.

The wings should be removed by cutting as close to the shoulder as possible, severing them at the joints. Some prefer to remove the outermost tip of the wing while retaining the two inner shanks.

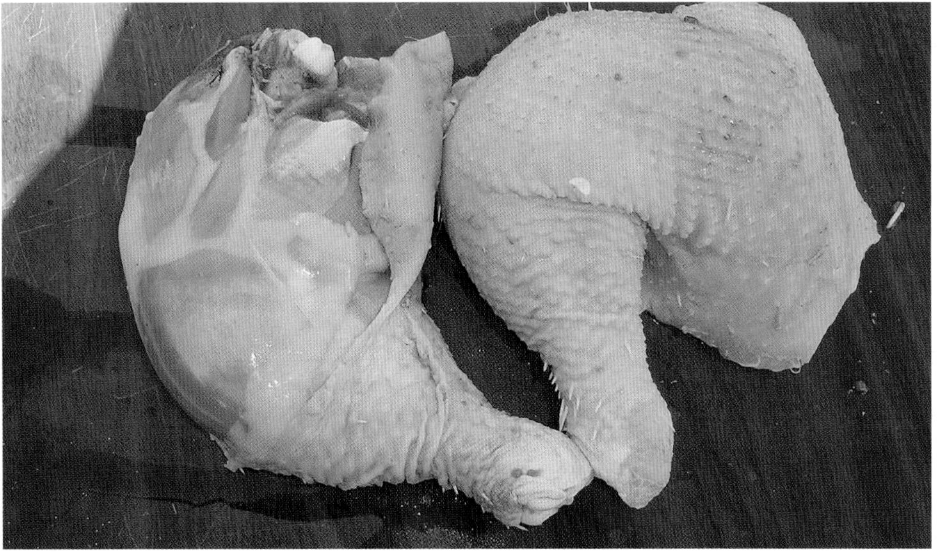

Your cuts should leave you with two legs and two breast pieces, as well as assorted internal parts. The heart and liver can be ground up to be used in sausage or dressing. After your cuts are finished, place those in cold water until you package them.

Remove shanks. Remove the feet and place the bird breast up on a table or a stable cutting surface while holding a shank (leg) in one hand.

Remove oil gland. With the bird breast down on the table, you can remove the oil gland near the tail. The oil gland can be seen at the base of the tail. Make a forward cut 1 inch from the gland and cut deep into the tail vertebra, then follow it to the end of the tail in a scooping motion to remove the gland.

Open body cavity. The body cavity can be opened by making a small cut near the rectum while being careful not to cut the intestines or contaminate the carcass with fecal material. Two types of cuts can be used to make this opening, depending how you will use the bird. A midline vertical, or J, cut is

Though not nearly as difficult as processing larger animals, it's still not an easy task to turn a live bird into dinner.

often used for broilers and other small poultry not to be trussed when cooked. The traverse or bar cut can be used for turkeys, capons, or other large fowl.

Then wash the inside of the carcass thoroughly with clean, cold water after you have finished removing the insides. The carcass is now ready for cooling.

Cooling. The carcass should be cooled as soon as possible after evisceration is complete. Place it in a cold bath of clean water at a temperature between 35 and 40°F to cool it quickly. After it is cooled, it is ready for cooking, freezing, or cutting it up further.

Most of the pieces are made by cuts at certain joints. Breaking down the bird into parts is a simple procedure. Typically, the edible yield for fryers and broilers is about 65 percent with the rest lost as bones and viscera. To cut up the carcass remove the wings and legs first, followed by the tail, and finally, the breasts.

The ribs may or may not be removed. You can split the breast lengthwise by first placing it skin side down. Then cut through the white cartilage at the

V of the neck. You can bend each side back as you push up on the breast from the bottom to snap the breastbone free. The wishbone is the clavicle and can be removed by severing it from the breast. Make your cut halfway between the front part of the backbone to a point where the clavicle joins the shoulder.

Meat Preservation

When butchering domestic animals, you likely will have more meat available than you or your family can eat at one meal. The rest will

There are hundreds of different ways to prepare sausage.

have to be preserved with one or more methods to keep it usable for later. Food preservation, in relation to meat products, is the process of handling harvested meat in a way that stops or retards the growth of microorganisms, making it safe for long-term consumption. The many forms of meat preservation include freezing, canning, drying, pickling, and curing and smoking. The successful harvest of home-grown animals and the proper preservation of the meat will greatly add to any farming experience you wish to have.

Methods

Freezing is one of the most commonly used preservative methods and has several advantages. It is a fast and simple way to stop microbial growth. When processing a carcass, it is important to remember that meat temperatures must be brought down to 40°F within four hours to prevent growth of spoilage microorganisms that lie deep within the carcass tissues or in the centers of containers of warm meat. If the meat has not been cooled but is going directly to a freezer, it must reach a temperature of 0°F within 72 hours to prevent the growth of putrefying bacteria.

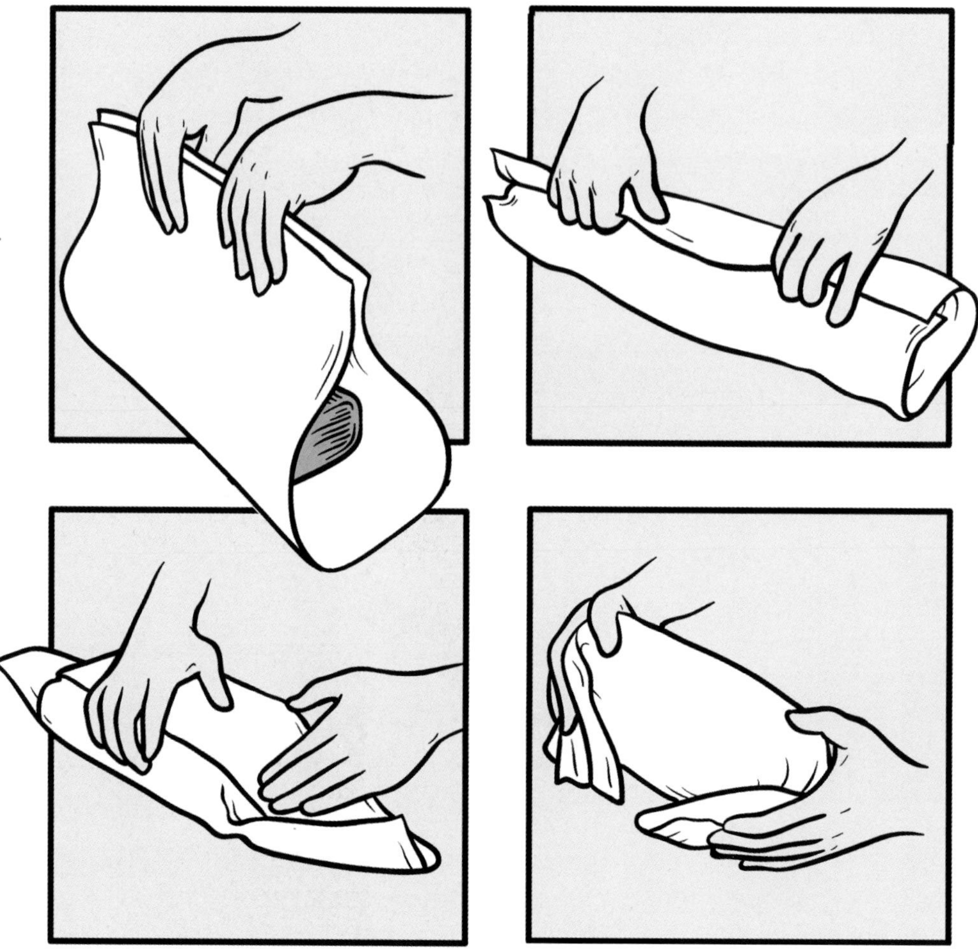

A moisture- or vapor-proof freezer wrap that seals out air and locks in moisture will make material to use for meat to be stored in a freezer. Choose heavy-duty, pliable wraps, such as freezer paper, aluminum foil, and freezer bags that can be used with bulky or irregular-shaped cuts of meat.

You should make cuts to be frozen into smaller, individual sizes that are ready for cooking rather than freeze them as large portions that need to be further deconstructed once they are thawed. Smaller cuts freeze more quickly and more evenly than very large pieces or chunks. It is better to minimize the number of times the meat needs to be handled and exposed to surfaces after it is thawed and used.

Canning involves cooking food to boiling point for a specified time as a form of sterilization. This is done while sealed cans or jars are submerged in boiling water or placed in a pressure cooker. Canning is the second most commonly used preservation method for long-term storage of meat. Canned meats are generally of two types: sterilized and pasteurized. Sterilized meat products do not need refrigeration and can sit on shelves for extended periods as long as the container remains intact. Pasteurized products require refrigeration to inhibit spoilage.

Canning involves a time-temperature relationship in destroying most microorganisms. A specific internal temperature must be reached and held for a minimum amount of time to destroy the microorganisms present. A safe cook, which is considered to be one that destroys the botulism organisms, requires a minimum of three minutes at 250°F. Achieving this sterilizing temperature will require the use of a pressure cooker. These typically operate under pressure of 12 to 15 psi. Pressure changes the boiling point of water and allows it to rise above the normal boiling point of 212°F.

Drying is perhaps the oldest method of food preservation and involves dehydration of the meat. The removal of water from the meat significantly helps prevent, inhibit, or delay bacterial growth.

Pickling is the use of a brine, vinegar, or any spicy edible solution that is used to inhibit microbial action. It can involve two different forms including chemical pickling, which uses a brine solution, or fermentation pickling, which is used to make sauerkraut.

The term **curing** is sometimes interpreted to mean both curing and the subsequent smoking of meat. But curing is not smoking, although these two processes work together.

The curing process can be likened to a race between the penetration of the preserving salts and the produc-

Glass jars are used for canning meats and vegetables. Inspect all jars for rim cracks or chips, and discard any jars that have them because they will not create a good, safe seal. Also, inspect each rubber ring or metal lid with gaskets and discard any that are defective.

Hams should be rinsed or soaked in cold water before being cured or smoked. This removes excess salt on the outside and eliminates the formation of salt streaks on the meat when exposed to the heat of the smoker.

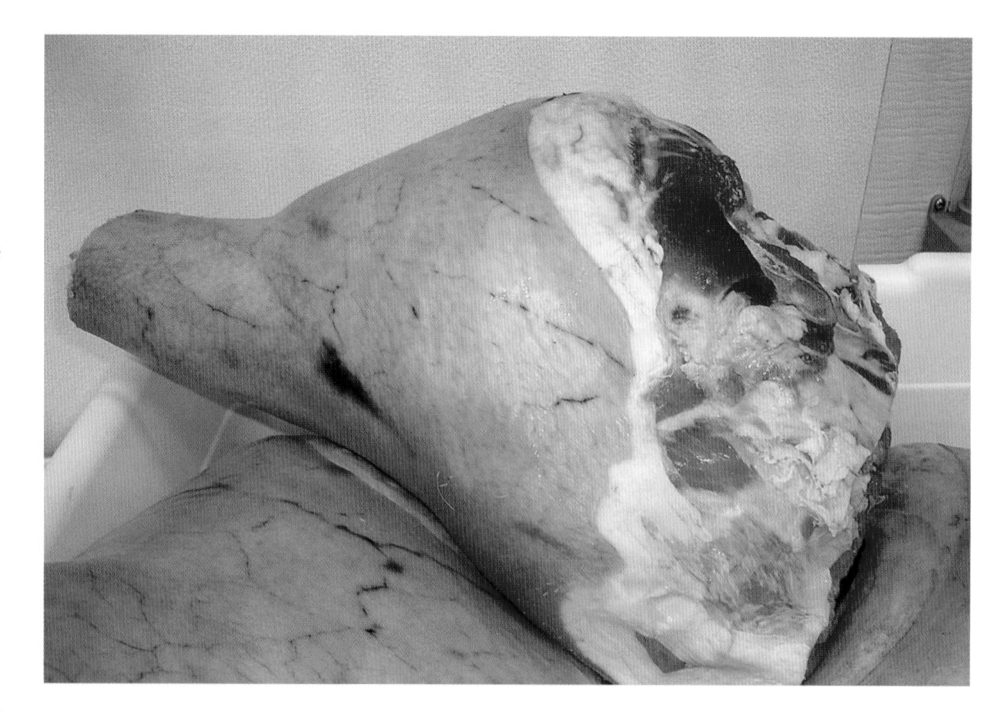

Dried beef is generally derived from extremely lean round or sirloin cuts found in the rear quarters of beef. These are less tender cuts because of the high amount of connective tissue found in them. When fully cooked, smoked, and thinly sliced, dried beef is an excellent meat for sandwiches. Although relatively dry, dried beef products should still be refrigerated.

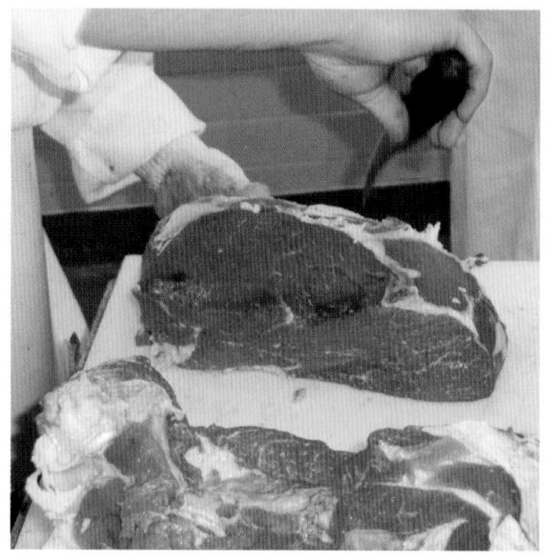

A barrel or wood smoker is easy to construct and can be used for small amounts of meat, fowl, and fish. Use metal racks to place or suspend your sausages. Wood is held in one part, and the smoke transfers to the large chamber. Set a container of water inside the large chamber to slow the drying process.

tion of spoilage bacteria in the meats. If you are considering home curing of your meat products, this is a race you must win to produce a safe food product.

The purpose of smoking meats is to give them a unique smoke flavor, an even external color, and a lowered moisture content, which reduces the opportunity for bacterial growth. Three factors affect the amount of time a meat product needs to be cured: the type of meat product, the density of smoke generated within the smoking unit, and the ability of the meat surface to absorb smoke.

The products used for creating the smoke will have a significant effect on the meat. Several options are available for smoking meat including natural wood smoke, charcoal, or electric units specially designed for smoking meats.

Natural wood smoke is generally produced from hardwood sawdust, wood chips, or logs. Hickory wood is the most popular for use in smoking although other hardwoods such as oak, maple, ash, mesquite, apple, cherry, and other fruit woods are also used. You should avoid using pine and other coniferous trees because of their high tar content and bitter flavor.

Sausages

You can make several types of sausages, including fresh, cooked and smoked, dry, and semidry.

Fresh sausages are made from uncooked and uncured cuts of meat and must be stored in a refrigerated or frozen state before being eaten.

Cooked sausages are usually made from fresh meats that are cured during processing, fully cooked and/or smoked. Cooked sausages should be refrigerated until eaten.

Dry and semi-dry sausages are made from fresh meats that are ground, seasoned, and cured during processing. They are stuffed into either natural or synthetic casings, fermented, often smoked, and carefully air-dried. True dry sausages are generally not cooked and may require long drying periods of between 21 to 90 days, depending on their diameter.

Fresh sausage is relatively easy to make, but must be handled carefully and cooked thoroughly.

TOP: Dry sausages have been cured, sometimes smoked, and dried, sometimes for months.

RIGHT: If you're going to use your fresh sausage within a day or two, you don't need to put it in a casing: just form it into patties and cook it.

It is not necessary to stuff fresh sausage meat into a casing. It can be left in bulk form or made into patties. But if ground into bulk form it will have to be used within one or two days to retain its freshness and quality. Most sausages are made by inserting the ground ingredients into some forming material that gives them shape and size and holds the meat together for cooking and smoking, or both. This material is called a casing.

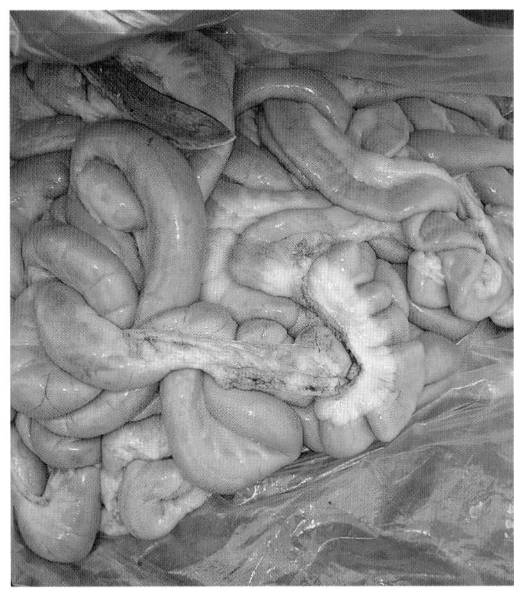

LEFT: Artificial or manufactured casings come in several sizes, thicknesses, and color for different types of sausages. Most manufactured casings must be soaked in clean water before use to make them pliable.

RIGHT: Natural casings are made from animal intestines, particularly sheep, pork, and beef. When using intestines for casings, they must be thoroughly cleaned and washed. The intestinal membranes are strong, flexible, and resilient, but they tend to lack the uniformity of manufactured casings.

Two types of casings are used in sausage making: natural and manufactured (also called synthetic). Although their purposes are the same, their origin is very different.

Natural sausage casings are made from parts of the alimentary canal of various animals that can include the intestinal tracts from pigs, cows, goats, or sheep. One advantage for using them is that they are made up largely of collagen, a fibrous protein, whose unique characteristic is variable permeability. This allows smoke and heat to penetrate during the curing process without contributing undesirable flavors to the meat. Natural casings can come from an animal that you are butchering. If using your own animal casings, it is important they are thoroughly flushed and cleaned and placed in a salt brine prior to use.

You can clean your own hog, goat, or sheep casings for sausage production after they are removed from the body cavity. Because they are unlikely to be the first parts you work with from the carcass, they need to be set in cold water to reduce their temperature to prevent spoilage. If you are working alone, set up the cold water tub prior to butchering. You'll need to consider several things if you are using the intestines for meat casings. The first is sanitation. The intestines will likely be filled with excrement typically containing *E. coli* bacteria.

Pig intestines make very good natural casings for your sausages because they are largely collagen and will easily break down during the curing process yet still are strong enough to hold the meat during the stuffing process. Their flexibility makes them an attractive alternative to synthetic casings. But good casings are also clean casings and you will need to prep them for use by removing all of the excrement and intestinal linings before using them. Turn them inside out and wash them in a cold 0.5 percent chlorine solution. Use a soft-bristle brush to gently scrub the excess fat, connective tissue, and any residual foreign or fecal materials off it. Be careful not to

Meat grinders reduce large pieces into a soft, pliable mass into which spices, fats, or other additives can more easily be mixed. Grinders can be hand- or electric-driven and come in many different designs and shapes.

overdo it or you may leave little tears in the membrane that can rupture and break during the stuffing process.

After thoroughly cleaning the intestines, rinse it with clean, cold water and invert it back to its original form. Use a saturated salt solution (1 gram salt/2.8 ml water) for storage overnight.

Generally, the larger the animal butchered, the larger the size of the intestinal tract you will have. The kind of sausage you want to make may have some bearing on the size of casings you use. Typically goats and sheep are the smallest, followed by pigs, and then cattle. Sheep and goat casings are more delicate and can be used for hot dogs, frankfurters, and pork breakfast sausages.

You will need a few pieces of equipment to make sausage in your home. The three most important pieces of equipment include a meat grinder (but not if you only purchase bulk sausage meat), a sausage stuffer, and a thermometer. Other pieces that you may find useful include a mixing tub, scale, and a smoker if you want to do your own meat smoking and preservation. Sausage-making equipment is usually available from meat equipment supply companies.

You may want to keep other items on hand during your sausage processing. Three important pieces include measuring instruments such as a scale for weighing meats and other ingredients, measuring cups, and thermometer(s) to monitor and maintain appropriate temperatures during the processing and cooking of sausages.

Modern sausage stuffers rely on the same principle as the old style, but are made of aluminum or stainless steel. Be sure to thoroughly wash, sanitize, and rinse the stuffer and funnels or horns prior to and following each use and between different sausage-making sessions.

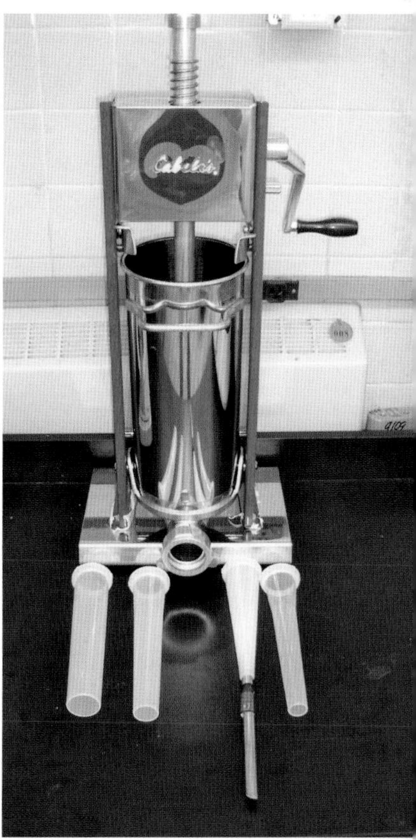

Chapter 7
Conservation

Protecting agricultural farmland needs to be one of the top priorities of farmers or anyone else involved in the ownership of wetlands, native grasslands, and other vital habitats. It is critical that these lands, which provide food for our tables and habitat for wildlife, are managed in a way that balances production with conservation.

Farmland, whether it be used for crops or livestock, must be managed in a way that ensures the continued health of the soil, water, and air.

Conservation and production can be partners in a long-range plan for your agricultural lands.

Doing so will help ensure that there will be productive farms and sufficient wildlife populations for future generations. We are the recipients of past generations' efforts to secure their futures, and we need to do the same for our descendants. It takes thousands of years to create fertile soils and only a few years to completely destroy them.

We must preserve our agricultural lands because we cannot create soil. We can compost, mulch, add fertilizers, chemicals and manures, but we cannot create more soil. Our efforts at preserving and protecting these productive lands are our best hope for the future.

Many conservation practices can be incorporated on your farm to help you do your part in this preservation while still using your land for production purposes. Conservation and production need not be separate components but can be partners in a long-range plan. Some benefits derived from this approach are environmental, including wildlife habitats, clean air and water, flood control, ground-water recharge, and decreasing carbon emissions.

Many county, state, and federal programs are available to assist you with developing conservation practices on your farm. One of the best sources of information about these programs is the Farm Service Agency, which administers the USDA programs at the county level. Check with your local FSA office for program sign-ups and eligibilities. They also have many

Use your livestock's manure to enrich your soil rather than pollute the waterways.

financial assistance programs that can help cover costs of conservation programs you wish to develop on your farm.

Practices that can help preserve, protect, and sustain your farmland include proper manure utilization (if you raise livestock), wildlife management, shelter belts, and woodlot management.

NUTRIENT MANAGEMENT

Handling and utilizing the manure produced by the animals on your farm in the most effective way is called nutrient management. Manure is a necessary byproduct of animals. How it is used once it is expelled from the animal will largely determine if it is a valuable asset to your farm or a costly liability.

The purpose of a nutrient management program is to use the manure components in a way that provides food for your soil through balanced applications across your fields, as well as minimizing any detrimental effects on the surrounding environment, such as waterways, streams, or other water sources, including wells.

With an increase in the number of farms with a large animal population, nutrient management has become a priority issue with farmers, environmentalists, government agencies assigned with the task of maintaining clean water resources, and the public at large.

Many problems encountered with manure result from runoff of these highly concentrated nutrients when it is spread on the top of fields. Most often runoff occurs in connection with frozen ground where the manure has

A tractor and tractor-drawn manure spreader are very useful pieces of equipment if you have a large number of animals. Handling manure in a sustainable way is a critical part of managing your farm.

little or no chance of being absorbed into the ground prior to rain or snow melt. Farm waste runoff into streams, creeks, rivers, lakes, and other water bodies has become one of the most contentious issues between rural landowners and urban populations that view these waterways as recreational areas. These contentious paths do not need to cross. Well-planned usage of animal waste products—manure—can avert problems before they occur and can be a valuable asset instead of a liability to your farm.

Manure Handling

The manure produced on your farm can be identified as three types: solid, semisolid, and liquid. Each of these types needs to be handled in different ways to make effective use of them.

Solid manure consists of a combination of fecal matter and urine that has been mixed with dry bedding materials, such as straw, hay, sawdust, wood shavings, corn fodder, or any other material that has become part of the bedding. Typically, solid manure is handled with a manure spreader that applies the material on croplands to be used as a fertilizer.

Piling manure on solid surfaces, such as cement, minimizes runoff and leaching before being spread. Having the flexibility of storing manure during times when it is difficult to get into a field will assist in planning a spreading schedule and avoid times of high runoff risk from rain or snowmelt.

Semisolid manure contains little or no bedding materials and has a consistency between liquid and solid, much like a thick pudding or slurry. This consistency creates some difficulty in storing it for spreading onto the fields. One solution is to use bedding and mixing it in to get a firmer consistency or to use a sloping floor to allow more of the liquid to drain to a holding area and then storing the solids to another area. By separating the parts, semisolid manure becomes easier to handle.

Liquid manure contains no bedding materials and is typically a combination of urine and feces. Containment of liquid manure requires a storage facility that does not let any of the nutrients leach into the ground, such as cement-lined pits or upright structures. In confinement buildings, a slotted floor system allows the manure to drop down into a pit. This type of process requires pumping the manure from the pits into tanks that can be taken to fields and spread on the open ground or injected into the soil.

Soil Conservation Plan

If you plan to participate in any federal farm programs, a soil conservation plan is required. The conservation plan is a part of any nutrient management program because it identifies crop rotations, the slopes of all fields, and the

Compost piles are an excellent method of handling stored solid manure in an environmentally sound way. Periodically stirring the compost pile will help transform the manure into more stable nutrients. In turn, this can be sold as mulch, garden manures, or other forms of fertilizer.

Organic and sustainable production systems have become popular with both farmers and consumers. Farmers concerned about the effects of chemical use on the soil, water, and wildlife have sought alternative methods of farming, such as pasture-grazing programs. Soil erosion is minimal where grass-based programs are used, especially when compared to row-crop farming.

Properly designed buffer strips along creeks and waterways provide a filter to prevent pasture or field runoff from entering streams. These buffers may be grazed for short periods of time to make use of the grasses lining the edges.

conservation measures you will need to follow to stay within the tolerable limits of soil erosion.

This plan also identifies which fields may have restrictions for spreading manure because of a close proximity to waterways, especially in the winter. One component of this plan includes identifying the best time of year when the manure-spreading can take place. This will depend upon the manure handling system on your farm. A farm with manure storage has a plan different from one that requires daily manure hauling. Many counties require a nutrient plan for any farm constructing new manure storage facilities or expanding its livestock operations.

Properly designed buffer strips along stream banks adjacent to fields with a potential for runoff can reduce the amount of manure entering a stream. These grassy strips help stop, filter, and hold back the sediments of runoff. Fencing plays an important role in buffer strips because it keeps cattle from overgrazing banks and walking in waterways. Nutrient and manure management plans and a conservation plan for your farm can be developed with help from your county agriculture extension agent or the Natural Resources Conservation Service.

Cattle and other farm animals will seek water to cool themselves during hot weather. Building buffer strips prevents animals from accessing the streams or waterways and prevents further pollution.

WILDLIFE MANAGEMENT

Wildlife management involves balancing the needs of the wildlife populations and your need to farm your land. They can be compatible goals, and you likely will be involved with custodial management practices that are preventive and protective rather than trying to manipulate populations through their introduction on your land.

Consultation with a wildlife management specialist will help determine the practices most appropriate for your farm land. They may be as simple as creating brush piles that are assembled to provide resting or escape cover for animals or birds. If they are adjacent to water sources, amphibians and reptiles may use them for breeding and resting.

You can establish permanent vegetation for wildlife by planting trees, shrubs, or flowers that may provide benefits for them. These will provide shelter and food at different times of the year. For example, planting wildflowers will provide nectar for bees, butterflies, moths, and hummingbirds and seeds for songbirds. Sowing cool-season grasses provides shelter for insects and rodents that attract wild grouse, wild turkeys, and gray fox, plus hunting sites for hawks, owls, and snakes.

Stream banks are important habitat for birds, fish, and other wildlife. Keep your livestock out of the stream and build buffers to absorb runoff before it reaches the stream.

Fencing stream banks, particularly if you raise livestock, is an important part of conservation. Livestock with free access to streams destroy wildlife and fish habitat, increase erosion and sediment production, and lower the water quality. Fencing stream banks keeps livestock away from sensitive water areas and encourages vegetation growth between the stream and fence. This area becomes a nesting site for birds and small mammals and a feeding place for myriad wildlife. A stabilized crossing can provide a limited area where livestock can drink from or cross the stream and minimize their damage.

Grass filter strips are usually found on cropland and consist of diverse grasses, wildflowers, or legumes. They help reduce runoff from fields into streams. The filter strip width is typically determined by the soil type, slope of the field, and type of pollutants found in the field. Filter strips can range in width from 20 feet from the edge of the waterway to 120 feet.

Contour grass strips and grass waterways can be created to provide channels for water runoff that does not create erosion. They can be constructed of perennial grass along cropland strips to prevent erosion, filter any runoff, provide wildlife cover and food, and reduce sediment disbursement.

Wildlife habitat buffers provide natural cover and food for bobwhite quail in cropland areas. Marginal pastures can be used to establish buffers as well as streambanks.

Wetland restoration can be done on lands that were previously drained, sometimes to make cropland, to restore them to their natural state as a wetland. Restoration projects may include removing tiling buried for drainage, filling ditches, or removing dams or levees designed to keep water out. These wetlands can provide breeding, nesting, and feeding habitats for amphibians, waterfowl, shorebirds, songbirds, and many other species.

Fallow land practices involve leaving a field or other area idle for a time and not using it for production. It is a rest period for the land because it does not expend many nutrients for crops and plants. Following a fallow land program does not alone make a suitable wildlife habitat management practice, but when used with other practices, such as rotational grazing, it can be useful.

SHELTERBELTS

Shelterbelts are windbreaks designed to protect farmlands and livestock from wind and blowing snow and promote wildlife habitat. Windbreaks work well in many cold climate areas because they allow the livestock and wildlife protection from the bitter wind. If the windbreaks are properly placed, they can provide benefits to feeding areas and pastures. Reducing the wind speed in winter lowers animal stress, increases feed efficiency, assists animal health, and makes the working environment for you a bit more pleasant.

The windbreak should be designed to meet the specific needs of your farm. There are two major types of windbreak designs: the traditional multirow design and a newer twin-row, high-density design. The design you choose depends on the area where you place it, the space available, and what you want the windbreak to accomplish.

Consider four basic factors when designing a shelterbelt: orientation of the shelterbelt in relation to the prevailing winds, width, plant arrangement, and species of plants selected.

Grassland birds benefit from pasture grazing. Mechanical harvesting discourages birds from nesting in fields where hay crops are mowed and taken off early in the season, but pastures where cows graze allow birds time to nest and raise their young without being mowed over. The insects scattered as the cows walk about are food for the birds.

Trees can serve as an effective windbreak for animals in cold winter conditions, especially if they are close to your buildings.

In most Midwestern states, northwest winds are the harshest in winter. Evergreen trees should be planted about 30 feet away from the north and west sides of a building; they are most effective if located about 150 feet upwind from areas to be protected. The base of the trees should be trimmed out to 10 to 20 percent of their height to keep snow from clogging the windbreak and the area immediately downwind.

The width of a shelterbelt can vary depending on the amount of space available. Ten or more rows are ideal but some benefits can be derived from just a single row of trees. Plant at least three rows of trees if possible with the center row trees staggered with those on the outside rows. The traditional windbreak can be made of three or more rows of trees, shrubs, or tall poles placed in the ground with galvanized sheeting nailed between them. The twin-row, high-density design is more elaborate and utilizes closer spacing between the rows and the trees and shrubs in each row.

Use conifer trees and plant them facing the north or west, allowing 15 to 18 feet of space between spruces. The innermost rows can be made up of tall, deciduous trees to help divert the wind over the top of the area to be protected. Shrubs planted in the outermost rows will catch drifting snow. A shelterbelt consisting entirely of deciduous trees (those which drop their leaves in winter) is ineffective for control because it does not offer enough resistance to air passage.

To avoid tree kill, shelterbelts used near open livestock lots should not be in the path of lot drainage.

Trees provide a good windbreak, but they may take years of growth to reach maximum effectiveness. An alternative to trees is to construct a wooden or metal windbreak that can accomplish the same purpose.

WOODLOT MANAGEMENT

Decide what your goals are for your woodlot. You can have a forest that provides a steady supply of firewood and shelters wildlife at the same time.

Woodlot management can improve timber growth, create opportunities for outdoor recreation, and increase the value of your land. Well-managed stands of trees are more resistant to fire, wind, and disease and can promote new growth. Woodlot productivity is not

Talk with a forester to create a plan to regenerate your woodland. You may need to remove dead trees to make room for young seedlings.

measured by the wood you harvest now but by the amount of wood the land continuously produces.

Most small landowners will only sell timber once or a relatively few times in their lifetime. Economies of scale make logging on small tracts of land uneconomical.

For long-term firewood or pleasure, you may need to regenerate your woodland. You can regenerate in several ways, including removing dead trees not wanted for wildlife habitat, cutting poor-quality or full-grown trees to make space available for new seedlings, thinning young trees to promote faster growth, and planting new trees.

Careful tending, harvesting, and regeneration can create vigorous tree growth and wood production, as well as improving and encouraging wildlife habitats. The first step is to identify your ownership goals and objectives. Perhaps you want firewood, timber for future building plans, or privacy, recreation, or hunting. A forester can help you evaluate your goals and whether they are compatible with the resources on your property.

Glossary

abortion. Loss of pregnancy before going full term

abrasive grinding. Using a grinding wheel with a rough finish

acre. A unit of land measurement that equals 43,560 square feet

afterbirth. Placenta or fetal membranes expelled from the uterus after birth of young

aging. The time process involved causing a maturing or ripening of meat enzymes that increase flavor and has a tenderizing effect

aitch bone. The rump bone

alternator. An electric generator for producing alternating current

ampere. A measure of electrical current that indicates how many electrons are passing through a given point in the circuit

annual. A plant that only survives for one growing season and does not overwinter

anterior to. Toward the front of the carcass, or forward of

arc welding. A process that uses an electric current to generate heat needed to create a molten puddle of metal to form a weld

artificial insemination (AI). Process where semen is placed within a female's uterus by artificial means rather than through sexual intercourse

average daily gain. Amount of weight that an animal gains each day

backstrap. An inedible, yellowish-colored connective tissue composed of elastin running from the neck through the rib region of beef, veal, and lambs, and also the base of the ribs

bactericides. Any agent that destroys bacteria

bacterial wilt. A disease that invades the vascular (water conducting) tissues in leaves, causing a rapid wilt of the plant

bagging. The rapid expansion of the udder in anticipation of giving birth, beginning up to two weeks before

balanced ration. Food for animals that includes all the daily required nutrients

ballast. Weight added to the tractor to improve traction

Bang's disease. Common name for brucellosis, a bacteria that causes abortion

bantam. A small breed of chicken

barrow. A male pig that has been castrated or neutered, usually as a young pig

beef (noun). The meat of a cow, bull or steer

beef (verb). To send to slaughter or to butcher

blade meat. The lean meat overlying the rib eye and rib portion of the primal cut

bloat. Digestive disorder of ruminants, usually characterized by an excessive accumulation of gas in the rumen

block heater. A device attached to the engine that supplies heat to raise the temperature of the engine coolants to help it more easily start in cold weather

boar. Adult male pig kept for breeding purposes

bolt. A plant going to seed

bone-in cuts. Meat cuts that contain parts of bone

breech birth. Birth when a fetus is presented backward

broadfork. A two-handled fork up to 30 inches wide with 10 ½-inch tines; used for deep aeration in compacted soils

bruising. An injury that does not break the skin but causes discoloration in the muscle

buck. A male rabbit

bulk. The amount of physical space taken up by a food in relation to the nutrients it contains; often identified with hay and silages

bull. Male bovine

bulling. A slang term used to identify a cow in heat or estrous

butterfat. Fat content in the milk that can be separated out to make cream or butter

bushing. A metal sleeve that in the presence of grease serves as a bearing surface for a moving or rotating part

calipers. An instrument that measures small increments of scale

carburetor. A device that supplies the engine with vaporized fuel mixed with air

carcass. The dressed body of an animal slaughtered for food

carcass yield. Carcass weight as a percentage of live weight

carcass weight. The weight of the carcass after all initial butchering procedures have been completed and the internal organs, intestines, and fats have been removed

cereal. Plants of the grass family that yield an edible, starchy grain such as corn, wheat, oats, rye, barley, or rice

chick. A baby chicken

choke. A device that restricts the air intake passage to increase the fuel-to-air mixture

cleanings. A slang term identifying the expelled placenta after giving birth

clostridial disease. A deadly disease caused by spore-forming bacteria including tetanus, blackleg, malignant edema, and entertoxemia

coccidiosis. Intestinal disease and diarrhea caused by protozoans

cockerel. A male chicken under one year

collagen. A fibrous protein found in connective tissue, bone, and cartilage

colostrum. First milk after a female gives birth; contains antibodies that give the newborn temporary protection against certain diseases

colt. A male horse under the age of three

compost. A mixture consisting largely of decayed organic matter, such as dried leaves, hay, straw, grass clippings, eggshells, coffee grinds, vegetable peelings, as well as manure; used for fertilizing and conditioning soil

compressed air system. A method for creating high-pressure air for cleaning

conformation. Physical size and shape of an animal

creepers. A rolling table on which a worker can lie for access to undercarriages of equipment or vehicles

crop rotation. The deliberate alternation of plant locations

crossbreed. Animal whose parents are of two different breeds

culling. Process of removing animals that are below average in production or are physically unsound

cultivar. A named plant selected for growing because of some valuable attribute

cud. A wad of food that is regurgitated from the rumen to be rechewed

cure. Any process to preserve meats or fish by salting or smoking, which may be aided with preservative substances

cutting yield. The proportion of the weight that is a salable product after trimming and subdivision

dam. Female parent

differential. A device found in transmissions that redirects the power delivered by the engine to the individual rear drive axles in a way that applies the power differently, allowing the outside wheel, during a turn, to rotate more often than the inside wheel

direct-sowing (or direct seeding). Planting seeds directly into the garden rather than starting them indoors first

disc harrow. Implement with one or two rows of offset discs, often used in seedbed preparation

doe. A female rabbit

dorsal. Toward the back of the carcass; upper or top line

double-cropping. Planting another crop after the first one is harvested

downer. Slang term referring to an animal that can't return to its feet due to any cause

drawbar. A solid metal bar at the lower rear end of a tractor to which a piece of equipment is attached for pulling

dressing percentage. Percentage of the live animal that ends up as carcass

drones. Male bees

dual purpose breeds. Breeds used for more than one purpose; such as both meat and milk

ductility. The ability of metal to be shaped by heating or cooling

dwarf. A rabbit that matures to less than 4 pounds

electrolytes. Chemical salts that ionizes in solution. Ions increase the electrical conductivity of solutions, often affecting muscle movements

estrus. Period of time when a female is sexually receptive to a male; also referred to as heat

ewe. Female sheep

ewe lamb. Female sheep less than one year old

extractor. A large, stainless steel drum that is used for extracting honey from frames

fabrication. Process of cutting a whole carcass into smaller portions; in metal work refers to the construction of an object to be used

farrow. Refers to female swine giving birth

fatten. To feed for slaughter; make fleshy or plump

feed efficiency. The number of pounds of feed required by an animal to gain 1 pound in body weight

feedlot. Area where many animals are fed or finished for market

field crop. A crop such as wheat or corn, as opposed to a garden crop, such as tomatoes

fillet. To slice meat from bones or other cuts. Also refers to boneless slices of meat which form portion cuts

filly. A female horse under the age of three

fish emulsion. A high-nitrogen, organic fertilizer made from processed fish wastes

flight zone. Distance you can get to an animal before it flees

flock. Small group of sheep or poultry

foal. A baby horse

foliar feeding. The act of applying organic fertilizers, such as fish emulsion or kelp, directly to the plant leaves to be absorbed

forage. Pasture and field crops that are of value for diets that include leaves and stems

forequarter. The anterior portion of a carcass side such as the front shoulder

free-range. Chickens that are allowed to roam about in grassy areas

freemartin. A sexually imperfect female bovine that is sterile. This condition most often occurs when a female calf is born twin to a male

freezer burn. Discoloration of meat due to loss of moisture and oxidation in freezer-stored meats

freshen. Give birth to a calf and begin producing milk

fright or flight response. A behavioral reaction by animals to a stressful or threatening situation that increases heart, lung, and muscle activity

fryer. A rabbit, raised for meat, that is less than ten weeks old and weighs less than 5 pounds

gambrel. A frame used by those butchering for hanging carcasses

gasket. A piece of soft material sandwiched between stationary metal surfaces used to seal in oil and seal out dirt

gelding. A castrated male horse of any age

gestation. Pregnancy; fetus development period between fertilization and birth; in rabbits, 28 to 33 days; in horses, 340 days

governor. A device in tractor engines that is designed to regulate engine rpm. This is accomplished through automatic changes in throttle settings as needed in response to field conditions

grade. A designation that indicates quality or yield of meat based on standards set by the United States Department of Agriculture (USDA)

growing season. The length of time between frosts; the time between the last frost of spring until the first frost of fall

hands. A term used to describe the height of a horse; one hand equals 4 inches

hardened bolt. A fastener with a higher strength grade of metal used in its production to decrease breakage

hardware disease. Condition caused by a sharp foreign object penetrating the reticulum wall

heat. (See estrus.)

heirloom. Older, open-pollinated plant varieties that have been handed down from generation to generation

hen. A female chicken over one year

herbivore. An animal that feeds mainly on grass and plants

heritability. Likelihood of certain traits being passed genetically to future offspring

hindquarter. The posterior portion of the carcass side

hive. A structure to house bees, typically consisting of a hive stand, a landing board, a bottom board, hive boxes, supers, an inner cover, an outer cover, and an entrance reducer, as well as frames in which honey, brood, and wax are stored

horsepower. A unit of energy. Specifically, 1 horse power is the amount of energy required to lift 33,000 pounds 1 foot over a 1-minute time period

hybrid. The genetic crossing of two distinctly different plants, resulting in a plant with increased hardiness and disease resistance but lacking in the ability to reproduce itself

hybrid vigor. Increase in the performance resulting from crossbreeding

hydrometer. An instrument used to measure a liquid concentration

inbreeding. Mating of individuals more closely related than the average of the breed

intercropping. Planting more than one crop in a particular area

intoxication. When microbes produce a toxin that is subsequently eaten and sickness results in humans

intramuscular. Injection site into the muscle

intravenous. Into the vein; often used to quickly infuse fluids into an animal's circulatory system

jack. A male donkey

jenny. A female donkey

jumper cables. Wires used to transfer electrical power from one battery to another to aid in starting

junior. Any rabbit less than six months of age

ketosis. Incomplete metabolism of fatty acids, usually from carbohydrate deficiency or inadequate use of it; commonly seen in high-fat diets of cattle but can also affect other species

kindling. The birth and delivery of baby rabbits

kits. Baby rabbits not yet weaned from their mother

leaching. The downward percolation of minerals and organic nutrients through the soil

legumes. Plants that are members of the trifoliate family because their leaves come in threes; includes soybeans, peas, alfalfa, and clovers

length of carcass. A measurement taken on the carcass or is estimated on the animal; the distance from the front edge of the first rib to the front end of the aitch bone

lethal defect. Death of a fetus or animal caused by its genes

loineye area. The number of square inches in a cross-section of loineye muscle; taken by cutting the loin between the 10th and 11th ribs and measuring the cut surface area of the muscle

lugs. That portion of rear steel wheels that are bolted to the outer circumference and provide traction

magneto. A device that uses the principle of electromagnetic interactions between magnets and wires to create a spark in the combustion cylinder of an engine at the proper time

manifold. That part on an engine that directs the intake air and exhaust in and out of the engine

marbling. Streaks and flecks of fat located within the muscle or between muscle groups

mare. A female horse over the age of three

market weight. Weight of the animal when sold for processing

mastitis. An infection of the mammary gland caused by bacteria

maternal traits. Characteristics of a good mother or those lacking

metritis. Inflammation of the uterus

peritonitis. Inflammation of the lining of the abdominal cavity; usually results from a puncture wound or hardware that pierces the stomach wall and fermentation fluids enter the space between the body wall and internal organs

MIG. Metal inert gas welding, or GMAW, gas metal arc welding; process that allows very fine control of heat applied to the weld, which allows for very thin pieces of metal, such as sheet metal, to be welded together

milling. Term used in metal work to describe cutting or grinding of metal

sandblasting. The process of using compressed air with granular particles to remove rust, paint or corrosion

moldboard plow. A metal blade that turns soil over making it ready for further seedbed preparation

mulch. A light layer of straw, hay, wood shavings, or grass clippings

mule. The offspring of a male donkey and a female horse

muscle pH. The acidity or alkaline level in the muscle; generally declines after harvest and the rate of decline is an important factor affecting meat quality

mycotoxin. Toxic substance found in molds

NFE (narrow front end). A style of tractor that has two front wheels mounted very closely together, giving it the appearance of a tricycle

open. An animal of breeding age that is not pregnant or has not been bred

open pollinated. Any plant pollinated by wind, insects, or other natural means; nonhybrids; reproduce true from seed and are virtually identical to the parent plant

organic farming. Holistic, ecologically balanced agriculture that meets the USDA National Organic Standards

outcrossing. Mating of individuals less closely related than the breed average

paddock. Enclosed area for grazing animals

palatable. A feed that tastes good to and is easily chewed by the animal

parturition. Act of giving birth

pasteurize. To heat milk to a certain temperature to kill bacteria present

perennial. A plant that survives for more than two growing seasons

perimysium. Connective tissue covering and binding together bundles of muscle fibers

petcock. A small faucet or valve of an engine for releasing a gas or air, or draining

pinkeye. Condition where the membranes lining the eyelids and eye covering become infected

pneumonia. Infection of the lungs

poll. Top portion of the head of an animal

polled. Animal that naturally has no horns

pony. An equine that is less than 14.2 hands

posterior. Toward the rear of the carcass; behind

pressure canning. The process used for canning low-acid foods, such as vegetables or meat

price spread. Difference between the farm and retail prices

primal cuts. Basic major cuts into which whole carcasses and sides are first separated

power takeoff (PTO). An engine- or transmission-driven shaft protruding at the rear of a tractor that transfers engine power to the trailing equipment

pullet. A female chicken under one year

purebred. Animal descended from a line of ancestors of the same breed; may or may not be registered with an association

queen bee. The only egg-laying female in the hive

raised beds. Garden beds whose soil level is raised above the surrounding soil

ram. Intact male sheep

ration. Amount of feed eaten or provided within a 24-hour period

raw milk. Nonpasteurized milk; milk straight from the lactating female

registered. Purebred animals whose pedigrees are recorded in a breed registry

render. To melt down the fat

resistor. An electrical device that limits the flow of electricity

retained placenta. Failing to shed the placenta after giving birth

rooster. A male chicken over one year

ROPS (roll-over protection systems). Systems designed to protect the operator should the tractor overturn; comprises a roll-bar or cab, and a seat belt

rotational grazing. Program where animals are moved from one pasture area to another

row crop. A style of tractor whose wheels are designed to straddle crops grown in rows

scours. A term used for diarrhea, most often in young offspring

scurs. Small, rounded portions of horn tissue attached to the skin at the horn pits of polled animals

senior. Any rabbit over six months of age; in some breeds (known as 6/8 breeds), any rabbit over eight months of age

settle. Slang term used to identify an animal that conceived from breeding; failing to conceive from a mating or insemination is referred to as not settling

shear bolts. Bolts that offer protection to equipment by breaking when sufficient resistance is applied from a counter force

sheet metal. Any flat, thin layer of metal used in fabrication

shrinkage. Weight loss that may occur throughout the processing sequence; sometimes due to moisture or tissue loss from both the fresh and the processed product

side. One matched forequarter and hindquarter, or one-half of a meat animal carcass

silage. Crop that has been turned into animal feed through fermentation

silverskin. The thin, white, opaque layer of connective tissue found on certain cuts of meats; usually inedible

siphon tank. A device used to pull out fluids from reservoirs using a suction action

sire. Male parent

skimmed milk. Milk from which most of the butterfat has been removed

slow-moving vehicle sign (SMV). A triangular, florescent-colored caution sign placed on the rear of a tractor or equipment that travels at a significantly reduced speed from normal traffic on highways

soldering. The process of using low heat to join soft metals together with low melting points

solids. The nonwater ingredients of milk that include fat, proteins, and minerals

solids-nonfat. The total solids in milk without the butterfat protein

springer. A cow or heifer that is about to calve

stallion. A male horse over the age of three

standing heat. A period of time during estrus when the female will stand still while being mounted by another female or male

steer. A castrated male bovine

stillborn. Offspring that is not alive at birth

stocking density. Relationship between the number of animals and area of land available for grazing

stocking rate. Number of animals grazing a unit of land for a specified period of time

sub-primal cuts. The subdivisions of the wholesale or primal cuts that are made to make handling easier and reduce the variability within a single cut

succession planting. Planting seeds continually throughout the season to ensure a continual crop

supplement. Any addition to a balanced ration that may be needed; often vitamins or minerals

sustainable farming. The process of balancing what is taken out of the soil and what is returned to it without relying on outside products

tattoo. Permanent mark in the ear often signifying vaccination for brucellosis

teat. The elongated protuberances on a female's udder where the milk exits; most correctly pronounced as teet

tempering. Heating metal to a temperature just below melting and letting it cool to add strength

thermostat. An automatic device for regulating temperature in an engine

three-point hitch. The attachment points at the rear end of a tractor that are used for vertical lift of equipment; often two side points and one higher, center point

threshing. The process of removing the grain from the plant

torque wrench. A fastener that provides a way to measure the amount of twisting force (torque) being applied to a bolt being tightened

trio. A group of three rabbits, consisting of one buck and two does, all of the same breed; intended for breeding

trocar. A sharp-pointed instrument equipped with a cannula, used to puncture the wall of a body cavity to release gas or withdraw fluid; often used in cases involving bloat

udder. Mammary glands, including the teats

vaccinate. To administer a vaccine into an animal's body to stimulate production of antibodies and immunity

vaccine. A fluid containing killed or modified germs

waste milk. Milk that can be used for calves, pigs, sheep, goats, but not for human food; often associated with colostrum and milk that contains antibiotics after treating for mastitis

water-bath canning. The process typically used for canning high-acid foods, such as fruit

weaning. Removing the young from the mother to stop nursing

weanling. A foal that has been weaned (usually refers to foals between the age of six months and one year)

wether. Male sheep that has been castrated

WFE (wide front end). Any tractor with a front axle that widely spaces the front wheels

winnowing. The separation of the chaff and bits from the grain

Workplace Hazardous Materials Information System (WHMIS). An international system of symbols that provides health and safety information about controlled products, such as solvents and flammable materials

withdrawal time. The interval between administration of a drug and the time of legal slaughter for meat or sale of milk; also refers to the length of time it takes for a drug to disappear from the system of the treated animal

worker bee. Female bees that perform all of the tasks in the hive

yearling. Refers to an animal between one and two years of age

yield. The portion of the original weight that remains following any processing or handling procedure in the meat selling sequence; usually quoted in percentages and may be cited as shrinkage

zirk fitting. A small valve allowing grease, under pressure, to be admitted while preventing contaminants from entering; prevents grease from escaping back through it

zones. Refers to USDA plant hardiness areas; used to help determine if winters will be too cold for various plants to survive

Resources

CHAPTER 1: THE BUSINESS OF FARMING

Books

Organic Farming, by Peter V. Fossel

Online Resources

Ecological Farming Association
www.eco-farm.org/resources

Food Safety Risk Management
www.fda.gov/Food/FoodScienceResearch/RiskSafetyAssessment/

The Learning Store, University of Wisconsin Extension
www.learningstore.uwex.edu/Direct-Marketing-C12.aspx

The Midwest Organic and Sustainable Education Service (MOSES)
www.mosesorganic.org

National Center for Appropriate Technology
www.attra.ncat.org/organic.html

Organic Farming Research Foundation
www.ofrf.org/education/database

USDA Grading Standards
www.ams.usda.gov

USDA Market News
www.ams.usda.gov/AMSv1.0/Market News

Wisconsin Local Food Marketing Guide (Wisconsin Department of Agriculture, Trade and Consumer Protection)
http://datcp.wisconsin.gov

CHAPTER 2: CROPS

Books

The Beginner's Guide to Vegetable Gardening by Samantha Johnson and Daniel Johnson

Homegrown Whole Grains: Grow, Harvest, and Cook Wheat, Barley, Oats, Rice, Corn and More by Sara Pitzer

Small-Scale Haymaking by Spencer Yost

Magazines

The American Gardener magazine
www.ahs.org/gardening-resources/gardening-publications/the-american-gardener

Northern Gardener magazine
www.northerngardener.org/

Online Resources

The American Horticultural Society
www.ahs.org

Baker Creek Heirloom Seeds
www.rareseeds.com

National Gardening Association
www.garden.org

Seed Savers Exchange
www.seedsavers.org

Sustainable Seed Company
www.sustainableseedco.com

CHAPTER 3: LIVESTOCK

Books

The ARBA Standard of Perfection published by the American Rabbit Breeders Association

The Beginner's Guide to Beekeeping by Daniel Johnson and Samantha Jonson

Family Cow Handbook by Philip Hasheider

The Field Guide to Horses by Samantha Jonson and Daniel Johnson

The Field Guide to Rabbits by Samantha Johnson

Horse Breeds: 65 Popular Horse, Pony, and Draft Horse Breeds by Daniel Johnson and Samantha Johnson

How to Raise Cattle by Philip Hasheider

How to Raise Chickens by Christine Heinrichs

How to Raise Goats by Carol Amundson

How to Raise Horses by Samantha Johnson and Daniel Johnson

How to Raise Pigs by Philip Hasheider

How to Raise Rabbits by Samantha Johnson

How to Raise Sheep by Philip Hasheider

Small-Scale Haymaking by Spencer Yost

Magazines

American Bee Journal
www.americanbeejournal.com

Backyard Poultry
www.backyardpoultrymag.com/

Bee Culture
www.beeculture.com

Beef
www.beefmagazine.com/

Chickens
www.chickensmagazine.com

Domestic Rabbits
www.arba.net

Equus
www.equisearch.com/magazines/equus/

Farm and Livestock Magazine
www.farmandlivestock.com

Goat Rancher
www.goatrancher.com

Hoard's Dairyman
www.hoards.com/

Horse and Rider
www.equisearch.com/magazines/horse-and-rider/

Horse Illustrated
www.horseillustrated.com

Pork Network
www.porknetwork.com/

Rabbits USA
www.smallanimalchannel.com/critter-magazines/rabbits-usa/2013-rabbits-usa.aspx

sheep!: The Voice of the Independent Flockmaster
www.sheepmagazine.com/

Online Resources

American Beekeeping Federation
www.abfnet.org

American Pastured Poultry Producers Association
www.apppa.org/

American Poultry Association
www.amerpoultryassn.com/

American Rabbit Breeders Association
www.arba.net

GrassWorks (resources for grazing)
www.grassworks.org

International Dairy Goat Registry
www.goat-idgr.com

The Livestock Conservancy
www.livestockconservancy.org

National Honey Board
www.honey.com

National Pork Producers Council
www.nppc.org

The United States Equestrian Federation
www.usef.org

CHAPTER 4: REPAIR AND MAINTAIN

Books

The Compact Tractor Bible by Graeme Quick

How to Keep Your Classic Tractor Alive by Spencer Yost

How to Keep Your Tractor Running by Rick Kubik

How to Set Up Your Farm Workshop by Rick Kubik

How to Use Implements on Your Small-Scale Farm by Rick Kubik

The Small Engine Handbook by Peter Hunn

Magazines

Gas Engine Magazine
www.gasenginemagazine.com

Vintage Tractor Digest
www.vintagetractordigest.com

Fencing Resources

Building an Electric Antipredator Fence, Oregon State University
http://ir.library.oregonstate.edu/xmlui/handle/1957/20669

Constructing Wire Fences, University of Missouri Extension Service www.muextension.missouri.edu/xplor/agguides

Construction of High-Tensile-Wire Fences, University of Florida
www.edis.ifas.ufl.edu/AE017

Everything You Need to Know About Electric Fences, Manitoba Agriculture, Food, and Rural Initiatives
www.gov.mb.ca/agriculture/livestock/beef/baa10s01.html

Fence for Deer Exclusion, USDA-National Wildlife Research Center
www.electrobraid.com/wildlife/reports/USDAMAY02.html

Fences for the Farm, University of Georgia
www.pubs.caes.uga.edu/caespubs/pubcd/c774.htm

Fencing Materials for Livestock Systems, Virginia Cooperative Extension
www.ext.vt.edu/pubs/bse/

Fence Planning for Horses, Penn State University
http://pubs.cas.psu.edu/freepubs/pdfs/ub037.pdf

How Windbreaks Work, University of Nebraska-Lincoln
http://nfs.unl.edu/documents/howwindbreakswork.pdf

Introduction to Paddock Design and Fencing-Water Systems for Controlling Grazing, National Sustainable Agriculture Information Service (ATTRA)
www.attra.org/attra-pub/paddock.html

Planning Fencing Systems for Intensive Grazing Management, University of Kentucky
http://www2.ca.uky.edu/agc/pubs/id/id74/id74.pdf

Seventeen Mistakes to Avoid with Electric Fencing, Sustainable Farming Connection
www.ibiblio.org/farming-connection/grazing/features/fencemis.htm

Temporary Fences for Rotational Grazing, Virginia Cooperative Extension
www.ext.vt.edu/pubs/ageng/442-130/442-130.html

Tractor Resources

Allis-Chalmers
www.allischalmers.com

American Society of Agricultural and Biological Engineers
www.asabe.org/publications/publications

International J.I. Case Heritage Foundation
www.caseheritage.com

John Deere Technical Information Book Store
http://techpubs.deere.com/deere/Default.aspx

Welding Resources

American Welding Society
www.aws.org

Miller Electric
www.millerwelds.com

Metals and How to Weld Them
by T. B. Jefferson

The Procedure Handbook of Arc Welding
by The James F. Lincoln Arc Welding Foundation

Safety in Welding, Cutting, and Allied Processes
American National Standards Institute (ANSI Z49.1); download at www.aws.org

Welders Handbook
by Richard Finch

Welding Complete
CPI

Welding Principles and Practices
by Sacks and Bonnart

CHAPTER 5: FARM SAFETY

Books

The Farm Safety Handbook
by Rick Kubik

Online Resources

Farm and Rural Safety, Canadian Provincial and Territorial Departments of Agriculture
www.agr.gc.ca/index_e.php

Farm Safety: Children
www.farmsafetyforjustkids.org

Farm Safety for Children: What Parents and Grandparents Should Know
www.ohioline.osu.edu/aex-fact/0991.html

Farm Safety Newsletter
www1.agric.gov.ab.ca/$department/newslett.nsf/homemain/far

International Society for Agricultural Safety and Health
www.isash.org

National Ag Safety Database
http://www.cdc.gov/niosh/topics/childag/default.html

CHAPTER 6: HARVEST, PRESERVE, AND BUTCHER

Books

The Ball Complete Book of Home Preserving by Judi Kingry and Lauren Devine

The Complete Book of Butchering, Smoking, Curing, and Sausage Making: How to Harvest Your Livestock & Wild Game
by Philip Hasheider

The Fresh Girl's Guide to Easy Canning and Preserving
by Ana Micka

Online Resources

Canning (food safety)
www1.extension.umn.edu/food/food-safety/preserving/canning/

Good Agricultural Practices Network
www.gaps.cornell.edu

National Center for Home Food Preservation
http://nchfp.uga.edu

Sample Farm Safety Manual
http://www.oregon.gov/ODA/ADMD/docs/pdf/gap_safety_program.pdf

Small grain growers
www.smallgrains.org

CHAPTER 7: CONSERVATION

Online Resources

Conservation Stewardship Program
www.sustainableagriculture.net/wp-content/uploads/2011/09/NSAC-Farmers-Guide-to-CSP-2011.pdf

Community supported agriculture (CSA)
www.nal.usda.gov/afsic/pubs/csa/csafarmer.shtml

Grow Biointensive (farming system)
www.growbiointensive.org

GENERAL FARMING RESOURCES

Books

Organic Farming, by Peter V. Fossel

You Can Farm: The Entrepreneur's Guide to Start & Succeed in a Farming Enterprise
by Joel Salatin

Magazines

Acres U.S.A
www.acresusa.com

FarmLife
www.myfarmlife.com/

Grit
www.grit.com/

Hobby Farm Home
www.hobbyfarmhome.com

Hobby Farms
www.hobbyfarms.com

Mother Earth News
www.motherearthnews.com/

Small Farm Today
www.smallfarmtoday.com/

Small Farmer's Journal
www.smallfarmersjournal.com/

Urban Farm
www.urbanfarmonline.com

Online Resources

Agricultural Innovation Center
www.fyi.uwex.edu/aic

AGWeb (powered by *Farm Journal*)
www.agweb.com/

Beginning Farmers
www.beginningfarmers.org

Beginning Farmers
Cornell Small Farms Program
www.smallfarms.cornell.edu/resources/beginning-farmer/

Building a Sustainable Business
University of Minnesota
http://www.misa.umn.edu/Publications/BuildingaSustainableBusiness/

Business Plan for Any Value-Added Agricultural Business
Wisconsin Department of Agriculture
http://datcp.wisconsin.gov/Business/Business_Resources/index.aspx

Building Sustainable Farms, Ranches and Communities
ATTRA
www.attra.ncat.org/guide/index.html

Crop mobs
www.cropmob.org

Enterprise Budgets
UW-Madison
www.cias.wisc.edu/category/economics/enterprise-budgets

Farm Information Resource Management
Michigan State University
www.firm.msue.msu.edu/

Farm Management Resource Guide
www.extension.iastate.edu/agdm/fieldstaff/resourceguide.pdf

Farm Service Agency
www.fsa.usda.gov/FSA/webapp?area=home&subject=fmlp&topic=bfl

The Farmer's Guide to Agricultural Credit
http://rafiusa.org/blog/the-farmers-guide-to-agricultural-credit/

Farmer Resource Network
http://www.farmaid.org/site/c.qlI5IhNVJsE/b.4375765/k.71EA/Farmer_Resource_Network.htm

Foreign Labor Certification
www.foreignlaborcert.doleta.gov

Hobby Farms resource page
www.hobbyfarms.com/localresources.aspx

Land Stewardship Project
Clearinghouse
www.landstewardshipproject.org/fb/resources.html

Lands of America
www.landsofamerica.com/america

Missouri Beginning Farmers
http://beginningfarmers.missouri.edu/

National Farmers
www.nfo.org

National Young Farmers Coalition
www.youngfarmers.org

Ohio State Farm Management
Ohio State University
www.aede.osu.edu/research/osu-farm-management

Penn State Extension: Start Farming
www.extension.psu.edu/business/start-farming

Risk Management Agency
www.rma.usda.gov/tools/agents/companies/RMA
www.rma.usda.gov/pubs/rme/fctsht.html

Start 2 Farm
www.start2farm.gov/new-to-farming

Strategies for Financing Beginning Farmers ATTRA
www.cfra.org/files/BF-Financing-Strategies.pdf

US Small Business Association
www.sba.gov/index.html

Photography credits

Peter V. Fossel: 16, 19, 20, 21, 22, 25, 27, 28, 29, 31, 61, 62, 64 from *Organic Farming* (Voyageur Press, 2007, 2014)

Philip Hasheider: 17, 26, 32 top, 81, 82, 83, 84, 121, 130, 140 bottom, 141, 283 from *How to Raise Pigs* (Voyageur Press, 2008, 2013); 69, 70 top, 71, 284, 285 bottom, 286, 287 *How to Raise Cattle* (Voyageur Press, 2007, 2013); 13, 32, 75, 76, 77, 122, 125, 133, 136, 139, 140 top *How to Raise Sheep* (Voyageur Press, 2009, 2013); 248, 250, 251, 252, 253, 254, 257, 259, 262, 265, 266, 267, 269, 272, 273, 274, 277, 278, 279 right from *The Complete Book of Butchering, Smoking, Curing, and Sausage Making: How to Harvest Your Livestock & Wild Game* (Voyageur Press, 2010)

Daniel Johnson/Fox Hill Photo: title page, 70 bottom, 115, 135, 219 top, 226, 241-246, 292 from *The Family Cow Handbook* (Voyageur Press, 2011); 23, 24, 108, 110 to 114 from *The Beginner's Guide to Beekeeping* (Voyageur Press, 2013); 38, 39, 40, 44 (right), 45, 46, 49, 51, 53, 54 from *The Beginner's Guide to Vegetable Gardening* (Voyageur Press, 2012); 55, 56, 85, 88 to 94, 117, 124 from *How to Raise Horses* (Voyageur Press, 2007, 2013); 86, 87 from *Horse Breeds: 65 Popular Horse, Pony, and Draft Horse Breeds* (Voyageur Press, 2008); 95, 96, 97 from *How to Raise Rabbits* (Voyageur Press, 2008, 2013); 98, 99 from *The Rabbit Book* (Voyageur Press, 2011); and dedication page, TOC, 8-9, 10, 11 all, 34, 35, 36 top, 37, 57, 58, 59, 60, 100 to 107, 120, 233, 234, 238, 281

Rick Kubik: 73, 128, 132, 134, 138, 219 bottom from *Farm Fences and Gates* (Voyageur Press, 2014); all photos on pages 170 to 186 from *How to Set Up Your Farm Workshop* (Voyageur Press, 2007); all photos on pages 72, 190 to 216, 221, 222, 224, 225, 227 from *The Farm Safety Handbook* (Voyageur Press, 2006)

Shutterstock: frontis (MaxyM), 12 (zschnepf), 15 (rebvt), 36 bottom (Charlotte Lake), 41 (Iakov Filimonov), 42 (Janez Habjanic), 43 top (ChameleonsEye), 43 bottom (Korolevskaya Nataliya), 44 left (motorolka), 47 (oksana2010), 48 (Leonid Shcheglov), 52 (Chris Bradshaw), 65 (Fotokostic), 66 (Catalin Petolea), 78 (Mayovskyy Andrew), 79 (Viachaslau Barysevich), 80 (Anna Azimi), 119 (Catalin Petolea), 126 (Julia Kuznetsova), 127 (Joy Brown), 187 (Fotokostic), 188 (Thomas M Perkins), 228 (jkelly), 229 (Catalin Petolea), 230 (mpetersheim), 232 (Zigzag Mountain Art), 235 top (TDMuldoon), 235 bottom (B Brown), 237 (Alena Brozova), 239 (Artography), 247 (Nancy Gill), 260 (margouillat photo), 261 (vlas2000), 264 (nito), 270 (MaraZe), 271 (B. and E. Dudzinscy), 275 (stocksolutions), 276 top (udra11), 276 bottom (Brittny), 279 left (aboikis), 280 (Catalin Petolea), 282 (foodonwhite), 288 (MaxyM), 290 (Becky Sheridan), 293 (anjun), 294 (MC_PP), 295 (MaxyM), 313 (ULKASTUDIO)

Spencer Yost: all photos on pages 142 to 153 from *How to Keep Your Classic Tractor Alive* (Voyageur Press, 2009)

Index

More great books from Voyageur Press

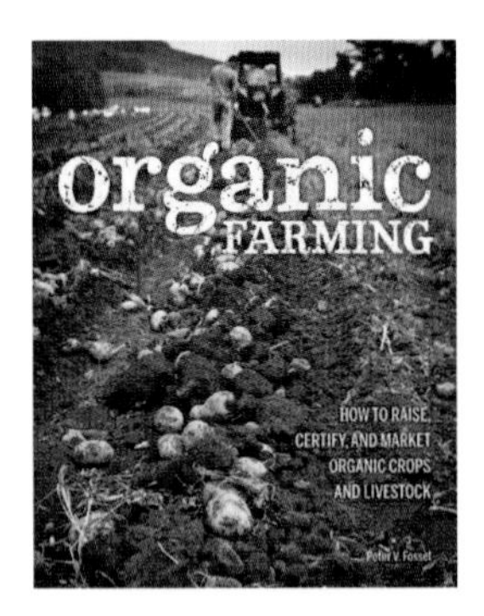

Organic Farming

ISBN:: 978-0-7603-4571-9

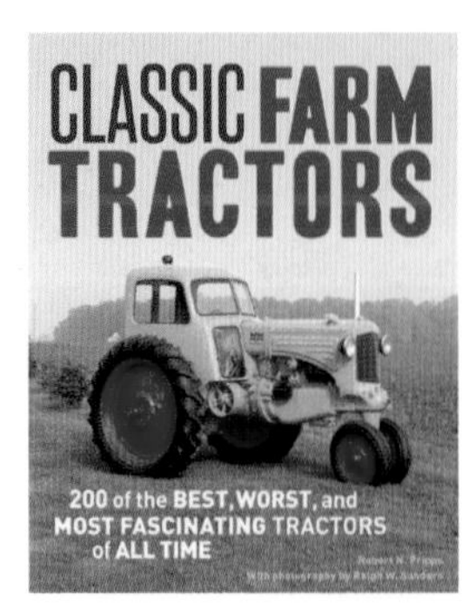

Classic Farm Tractors

ISBN: 978-0-7603-4551-1

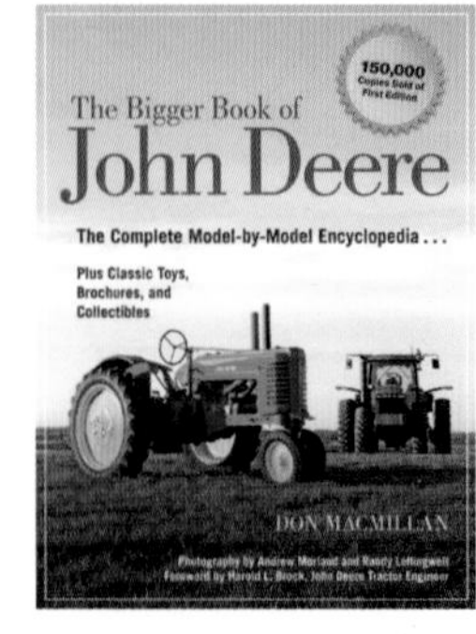

The Bigger Book of John Deere

ISBN: 978-0-7603-4594-8

Driving Horses

ISBN: 978-0-7603-4570-2

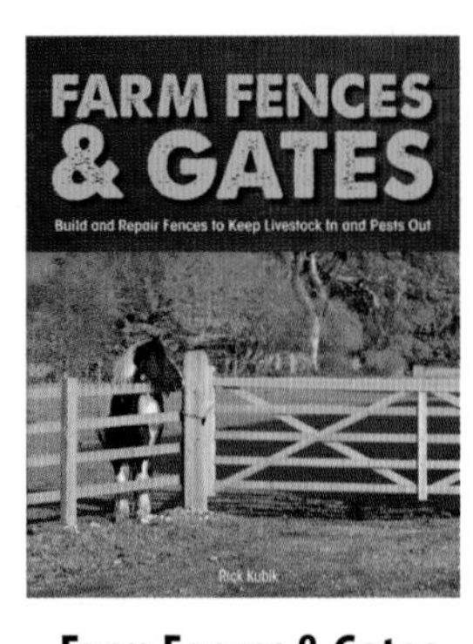

Farm Fences & Gates

ISBN: 978-0-7603-4569-6

The Beginner's Guide to Beekeeping

ISBN: 978-0-7603-4447-7

How To Raise Chickens

ISBN: 978-0-7603-4377-7

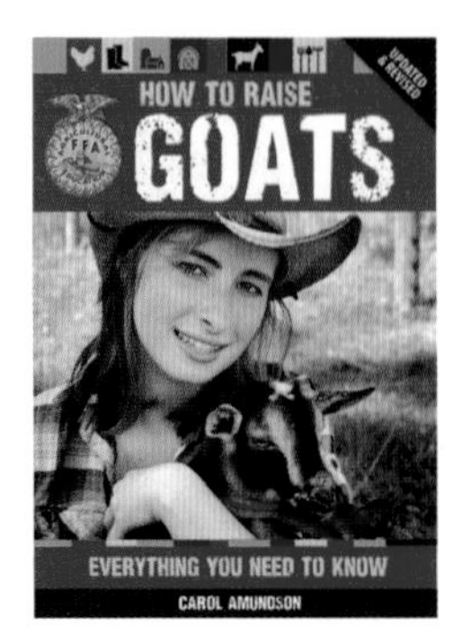

How To Raise Goats

ISBN: 978-0-7603-4378-4

The Whole Goat Handbook

ISBN: 978-0-7603-4236-7

WHOLE HOME NEWS

From the Experts at Cool Springs Press and Voyageur Press

For even more information on improving your own home or homestead, visit **www.wholehomenews.com** today! It's the one-stop spot for sharing questions and getting answers about the challenges of self-sufficient living.